Math Mammoth Kindergarten Worktext B

By Maria Miller

ISBN 978-1-954379-50-3

2026 Edition

Contents

Chapter 7: Shapes & Position Words

Chapter 8: More Addition and Subtraction

Foreword

Math Mammoth Kindergarten Workbook A and Kindergarten Workbook B comprise a complete math curriculum for the kindergarten mathematics studies. The curriculum meets the Common Core standards.

The main areas of study in these two workbooks are:

1. Basic counting skills within 0-100.
2. Addition and subtraction within 0-10, including word problems.
3. Identify, describe, analyze, and compose basic shapes.

Additional topics covered are sorting objects, comparing quantities and numbers, some foundations for measurement, position words, ordinal numbers, and counting coins.

This book, the kindergarten workbook B, covers shapes and measurement (chapters 5 and 7), two-digit numbers and coins (chapter 6), position words (chapter 7), and more addition and subtraction (chapter 8). The workbook A covers counting and writing numbers; groups, comparing, and sorting; addition and measurement; and subtraction. I heartily recommend that you read the full user guide in the following pages.

I wish you success in teaching math!

Maria Miller, the author

Foreword

User Guide

Note: You can also find the information that follows online, at **https://www.mathmammoth.com/userguides/** .

The Common Core Standards documentation is available at:
https://www.mathmammoth.com/preview/standards/MM-CCS-K-2026.pdf

Basic principles in using the Math Mammoth Kindergarten curriculum

Math Mammoth kindergarten curriculum is overall mastery-based, but it is slightly more spiral than the other grade levels of Math Mammoth, because most chapters deal with *two* major topics instead of one.

The kindergarten level differs from the other levels of Math Mammoth also in the fact that it is not expected that the child read the instructions or work independently very much. This level is meant to be **teacher-led**.

There are two student books (A and B). The chapters in them are intended to be studied mostly in order. However, you can do the lessons on shapes, measurement, and position words earlier or later, as you wish.

The **two student worktexts** (parts A and B) contain all the lesson material and exercises. Many of the lessons include activities that teach concepts, and those are explained right in the student books. Don't skip the activities, as they are an integral part of learning the concepts and skills for this level.

Some lessons include an explanation of the concept in a blue box. You, as the parent/teacher, need to explain the concept to the child(ren), using the blue teaching box as a guide. And then, almost all of the lessons contain some exercises for the child to do in the book.

Each chapter introduction explains the basic goals for the chapter. Most of them also list **games** (both online and offline games) that you can use to reinforce and practice the main skills in the chapter. Then, each chapter introduction also includes a link to a list of free online games and activities from other websites. All of these games can be used to supplement the math lessons, for learning math facts, or just for some fun.

We also include a versatile **worksheet maker** that allows you to make additional worksheets for the main topics in the curriculum. You will need Internet access to be able to use it. Access it here:

https://www.mathmammoth.com/private/Make_extra_worksheets_K.htm

Manipulatives

Below you will find a list of the recommended and optional manipulatives for the activities and games.

Recommended manipulatives

- a 100-bead abacus
- dice
- dominoes
- number cards, UNO cards, or regular playing cards
- a set of geometric solids (part B, chapter 7)
- a set of pattern blocks (part B, chapter 5)
- toothpicks and modeling clay (part B, chapter 7; for building geometric solids)
- a tangram set (part B, chapter 7)

Optional manipulatives

- counting bears (optional but can be fun because the sets often come with sorting cups and activity cards)
- math linking cubes (Unifix cubes)
- Cuisenaire rods (optional)

Many of the activities use various commonly found items that you probably have on hand in your home. For your convenience, we include a list of the recommended items and links to Amazon below. Note: You only need *one* item from each section. We're providing several options, plus naturally there exist many more choices for these products.

Recommended manipulatives

A 100-bead abacus or school abacus

There are lots of choices; search Amazon's collection here:
https://www.amazon.com/s?k=school+abacus&tag=mathmammoth-20

Learning Resources 10-Row Counting Frame or Abacus for Kids
This abacus has every five beads in alternating color (red, white), which is nice.
https://www.amazon.com/dp/B01LX6K2CV/?tag=mathmammoth-20

Wooden Abacus for Kids with Counting Sticks and Number Cards
https://www.amazon.com/dp/B08SBWKJZJ/?tag=mathmammoth-20

Geometric Pattern Blocks and Tangram-like games

Coogam Wooden Pattern Blocks Set 130PCS
https://www.amazon.com/dp/B07MYYK64R/?tag=mathmammoth-20

Beestech Wooden Tangram game with a puzzle Book
https://www.amazon.com/dp/B07Y37NNSR/?tag=mathmammoth-20

Vanmor Wooden Tangram set with shape cards
https://www.amazon.com/dp/B07XJ23KPG/?tag=mathmammoth-20

Wooden pattern blocks shape set with pattern cards
https://www.amazon.com/dp/B0CMTB3DR2/?tag=mathmammoth-20

GEMEM wooden pattern blocks set with design cards
https://www.amazon.com/dp/B07QPS5CFL/?tag=mathmammoth-20

Geometric solids (3-D shapes)

DEEKIN 36 Pcs Geometric Shapes $21
https://www.amazon.com/dp/B0BN8TXVK6/?tag=mathmammoth-20

DEEKIN 70 Pcs Wood Geometric Solids
https://www.amazon.com/gp/product/B0BC19RL8V/?tag=mathmammoth-20

Learning Resources Wood Geometric Solids Set of 19
https://www.amazon.com/dp/B000T1QQZ8/?tag=mathmammoth-20

Search Amazon's collection for geometric solids:
https://www.amazon.com/s?k=geometric+solids&tag=mathmammoth-20

Optional manipulatives

Cuisenaire Rods
https://www.amazon.com/dp/B016DU8UR0/?tag=mathmammoth-20

Counting Bears

Bmag counting bears
https://www.amazon.com/dp/B07ZNYDLXY/?tag=mathmammoth-20

Neorod counting bears
https://www.amazon.com/dp/B085ZSMK8P/?tag=mathmammoth-20

Linking Cubes

Learning Resources MathLink Cubes (100 pcs)
https://www.amazon.com/dp/B000URL296/?tag=mathmammoth-20

200 Pcs Math Linking Cubes
https://www.amazon.com/dp/B0GGCK8L8D/?tag=mathmammoth-20

Search Amazon's collection for math linking cubes:
https://www.amazon.com/s?k=math+linking+cubes&tag=mathmammoth-20

Pacing the curriculum

Each chapter introduction contains a suggested pacing guide for that chapter. You will see a summary on the right.

Most lessons are one page long, intended for one day. Some lessons are two pages, either intended for one day or two. There are also some optional lessons (e.g. about coins), not included in the charts on the right.

It can also be helpful to calculate a general guideline as to how many pages per week the student should cover in order to go through the curriculum in one school year.

Worktext A	
Chapter 1	19 days
Chapter 2	14 days
Chapter 3	25 days
Chapter 4	21 days
TOTAL	79 days

Worktext B	
Chapter 5	14 days
Chapter 6	17 days
Chapter 7	20 days
Chapter 8	13 days
TOTAL	64 days

The chart below lists how many pages there are for the student to finish in this particular grade level, and gives you a guideline for how many pages per day to finish, assuming a 170-day (34-week) school year. The page count in the table below *includes* the optional lessons (in particular, the lessons about coins).

Example:

Grade level	Lesson pages	School days	Pages to study per day	Pages to study per week
K-A	89	91	1	5
K-B	79	79	1	5
Kindergarten total	168	170		

The table below is for you to fill in. To get the count of "pages to study per day", **divide the lesson pages by the school days**. Lastly, multiply this number by 5 to get the approximate page count to cover in a week.

Grade level	Lesson pages	School days	Pages to study per day	Pages to study per week
K-A	89			
K-B	79			
Kindergarten total	168			

In general, kindergartners might spend 15-30 minutes a day on math, 1st-2nd graders might spend 25-40 minutes, third-fourth graders 30-60 minutes a day, and fifth-sixth graders 45-75 minutes a day. If your student finds math enjoyable, they can of course spend more time with it! However, it is not good to drag out the lessons on a regular basis, because that can then affect the student's attitude towards math.

Working space and the usage of additional paper

The curriculum generally includes working space directly on the page for students to work out the problems. However, feel free to let your students use extra paper when necessary. They can use it to draw diagrams and pictures to help organize their thoughts. Some students won't need the additional space (and may resist the thought of extra paper), while some will benefit from it. Use your discretion.

Concerning tests

While the rest of the grade levels in Math Mammoth include chapter tests, I feel that such testing is not necessary in kindergarten, as it can so easily lead to math anxiety and to viewing mathematics as a subject where correct answers are what matters most. Using tests has two sides. Even in other grade levels, the tests are optional. On the positive side, they can help the teacher know how their students are doing, and even help the student to know where they are at and to perhaps get some useful feedback. But all too often, testing leads to math anxiety. Also, students tend to study for the test only, just to forget the information right afterwards.

You can learn to assess your student(s) over the course of the year by observing them. This is even easier when you will be doing many activities with them that teach math concepts. You will be able to see where your child struggles and where they excel. You can even base a grade solely on your observations (if you need to give a grade). Or, you can use the chapter reviews as part of the grade.

The curriculum does include an optional end-of-year test, which is best used as a diagnostic or assessment test. If you do use it, it can tell you how well the child has mastered essential kindergarten math content.

Concerning the mixed reviews and the worksheet maker

The student books contain mixed review lessons which practice concepts and skills from earlier chapters. They allow you to spot areas that the student has not grasped well or has forgotten. When you find such a topic or concept, you have several options:

1. Check if the worksheet maker lets you make worksheets for that topic.
   https://www.mathmammoth.com/private/Make_extra_worksheets_K.htm
2. Check for any online games and resources in the Introduction part of the particular chapter in which this topic or concept was taught.
3. If you have the digital version, you can reprint a corresponding lesson from the worktext for the student.
4. Check if our online practice area at https://www.mathmammoth.com/practice/ has something for that topic.
5. Khan Academy has free online exercises, articles, and videos for most any math topic imaginable.

Frequently asked questions and contacting us

If you have more questions, please first check the FAQ at https://www.mathmammoth.com/faq-lightblue

If the FAQ does not cover your question, you can contact us https://www.mathmammoth.com/contact

Chapter 5: Shapes & Measurement
Introduction

In this chapter, we will study two-dimensional shapes and work on foundations for measurement again. We will encounter shapes again in Chapter 7. The lessons there will also include three-dimensional shapes. For now, the focus is on identifying basic shapes — no matter their orientation or size — and being able to count the sides and the corners each shape has.

The work in this chapter and chapter 7 lay groundwork for the student to describe basic attributes of shapes in first and second grades, which in turn prepares them to reason about shapes in a more formal manner later on. For example, in 4th and 5th grades, students will classify two-dimensional figures based on specific attributes related to angles and sides.

We have already looked at measuring length in the worktext part A. We review that topic, and then go on to learn about weight, in a very informal manner. Children don't need to actually measure the weight (or mass) of objects yet. They simply learn what "heavy" and "light" mean by comparing how heavy objects feel, and finding which of two objects is heavier (or lighter). Actually measuring the mass of objects will happen in second and third grades. However, feel free to let the child learn about the basic units for measuring weight/mass (pounds/kilograms) as you have opportunities, such as through everyday life.

Children also describe several measurable attributes of a single object. This means, for example, describing one object as both short and light, or as both tall and light, or as both long and heavy.

The lessons about times of day have to do with morning/afternoon/evening/night. They prepare the child to learn to tell time in first grade.

This chapter has quite a few activities in the lessons themselves. Don't skip them, as they are a major part of learning for the topics of this chapter. Children need to touch and feel the various shapes. They need to experience *light* and *heavy*; bookwork won't substitute for the actual hands-on experiences. You can also turn some of the activities into simple games.

Geometric Pattern Blocks and Tangram-Like Games

I recommend that you get a set of geometric 2-D shapes. This could be a set of pattern blocks or a tangram-like set of shapes. The latter usually come with design cards with composite shapes for the child to make.

Coogam Wooden Pattern Blocks Set 130 PCS
https://www.amazon.com/dp/B07MYYK64R/?tag=mathmammoth-20

Beestech Wooden Tangram game with a puzzle Book
https://www.amazon.com/dp/B07Y37NNSR/?tag=mathmammoth-20

Vanmor Wooden Tangram set with shape cards
https://www.amazon.com/dp/B07XJ23KPG/?tag=mathmammoth-20

Wooden pattern blocks shape set with pattern cards
https://www.amazon.com/dp/B0CMTB3DR2/?tag=mathmammoth-20

GEMEM wooden pattern blocks set with design cards
https://www.amazon.com/dp/B07QPS5CFL/?tag=mathmammoth-20

Pacing Suggestion for Chapter 5

Each lesson is intended to take one day.

The Lessons in Chapter 5	page	span	suggested pacing	your pacing
Basic Shapes 1	14	*1 page*	1 day	
Basic Shapes 2	15	*1 page*	1 day	
Drawing Shapes	16	*1 page*	1 day	
Shapes and Patterns	17	*1 page*	1 day	
Sorting Shapes	18	*1 page*	1 day	
Mixed Practice 1	19	*1 page*	1 day	
Measuring Review	20	*1 page*	1 day	
Weight	21	*1 page*	1 day	
Heavier and Lighter	22	*1 page*	1 day	
Times of Day	23	*1 page*	1 day	
More Practice with Measurable Attributes	24	*1 page*	1 day	
Times of Day 2	25	*1 page*	1 day	
Mixed Review Chapter 5	26	*1 page*	1 day	
Chapter 5 Review	27	*1 page*	1 day	
TOTALS		*14 pages*	14 days	

Games and Activities

This chapter has quite a few activities in the lessons themselves. You could use those activities on other days, too, not just during one lesson.

Collect the Shapes *(Identifying shapes)*

You need: A set of geometric shapes. Paper and pencil. One die. Assign the numbers 1-6 to specific shapes in the set, and write down this setup. For example, 1 = triangle, 2 = rectangle, 3 = hexagon, and so on. If the set does not have six different kinds of shapes, you can leave a number unassigned, or let some shape have two numbers assigned to it.

Game play: Spread the shapes in the middle of the players. On your turn, roll the die, and then find the shape that the number you rolled is assigned to. The winner is the person who first accumulates one of each kind of shape.

Problems with Patterns

You need: A set of geometric shapes.

Activity: This is a two-person activity or game. One person will use the shapes to arrange a pattern where the basic pattern is repeated at least three times. They will then mess up the pattern in one spot in some way, such as by replacing one of the shapes or removing one of the shapes. This is all done without the other person watching. Then the other person's task is to fix the pattern.

Games and Activities at Math Mammoth Practice Zone

Sort Spoons by Length — Online Practice
Sort spoons by dragging them on screen to the correct box. The instructions are provided as both text and audio.
https://www.mathmammoth.com/practice/sorting-game#questions=6&sort-by=length

Sort Shapes — Online Practice
Sort shapes by dragging them on screen to the correct box. The instructions are provided as both text and audio.
https://www.mathmammoth.com/practice/sorting-game#questions=6&sort-by=shape

Further Resources on the Internet

These resources match the topics in this chapter, and offer online practice, online games (occasionally, printable games), and interactive illustrations of math concepts. We heartily recommend you take a look. Many people love using these resources to supplement the bookwork, to illustrate a concept better, and for some fun. Enjoy!

https://links.mathmammoth.com/Kch5

Scan me

Basic Shapes 1

These are **circles**. They do not have any corners. They are perfectly round!

These are **squares**. They have *four* square corners and *four* sides that are the same length.

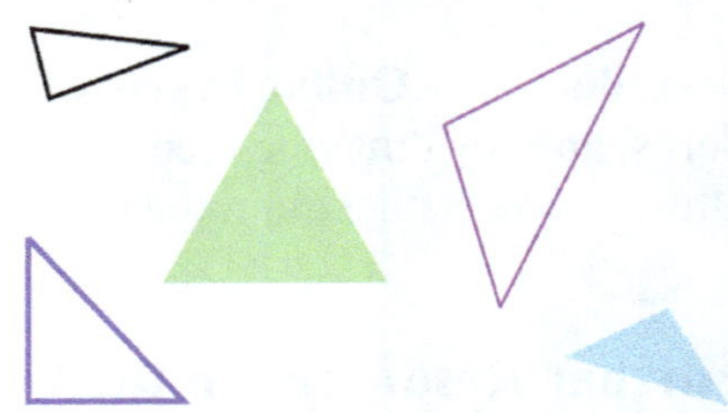

These are **triangles**. They have *three* corners and *three* sides.

1. **Activity:** Find all the squares among a set of pattern blocks. Then find all the triangles. Lastly, count the sides and the corners of one square and of one triangle.

2. **Activity:** Walk around your house, classroom, or some other location with your teacher, and look for objects that have the shape of a circle, square, or triangle. How many can you find?
   (For example, some lids are circle-shaped, some floor tiles are square-shaped, and some foods, such as chips and sandwich sections, are triangle-shaped.)

3. Write how many sides each shape has.

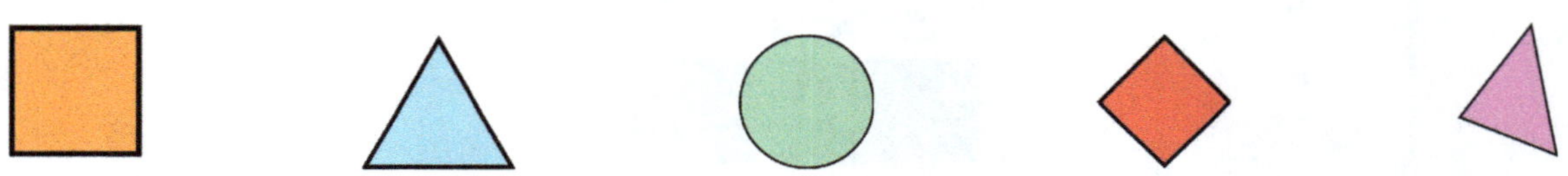

________ ________ ________ ________ ________

4. Color the shapes according to the color key.

circles ◯ = green

squares □ = orange

triangles △ = purple

5. Draw circles and color them to continue the pattern.

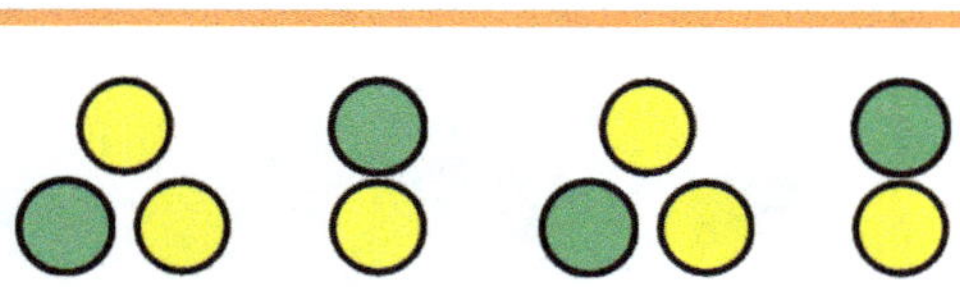

Basic Shapes 2

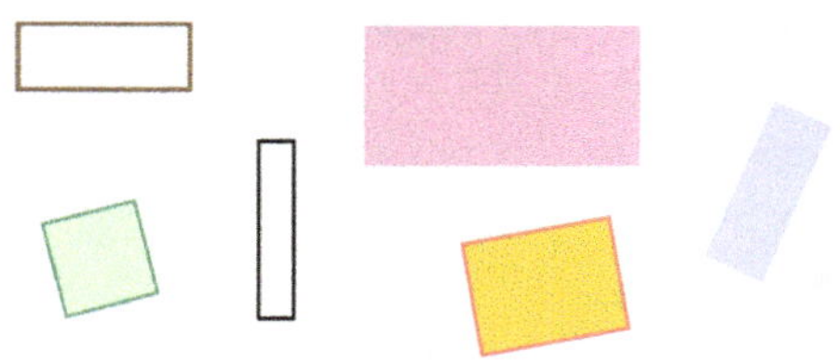

These are **rectangles**. They have four sides and four “square” corners. The sides don’t have to be the same length (though they may be).

These are **hexagons**. They have six sides and six corners.

1. **Activity:** Find all the rectangles among a set of pattern blocks. Then find all the hexagons. Lastly, count the sides and the corners of one rectangle and of one hexagon.

2. **Activity:** Can you find any objects in your surroundings that are shaped like rectangles or hexagons? Look and see!

3. Point at the shapes and say their names out loud to your teacher.

4. Circle the shape that has more sides. If they have the same number of sides, circle both.

a. 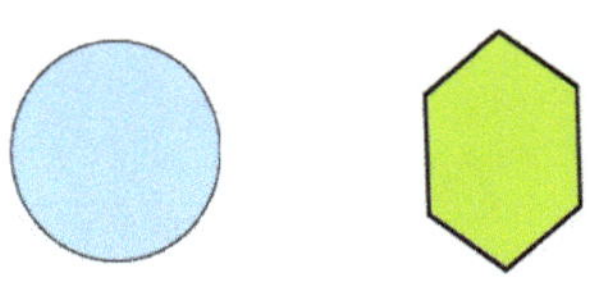

b. 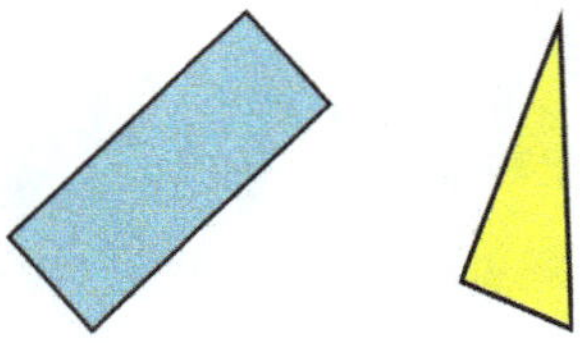

c. 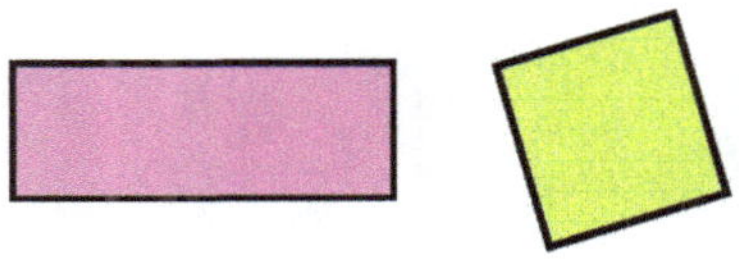

5. Write how many sides each shape has. Then color the shapes.

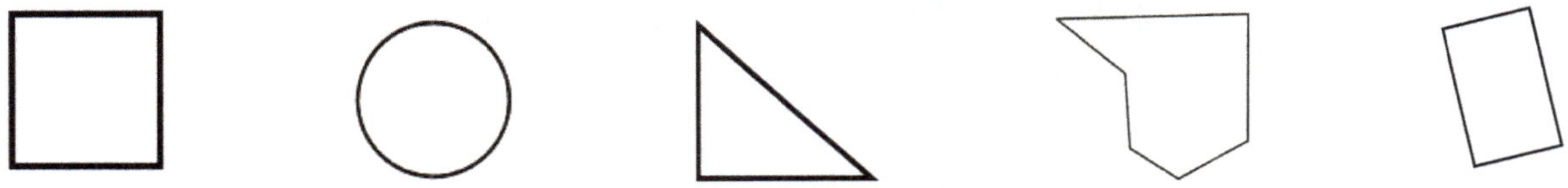

______ ______ ______ ______ ______

6. Counting on from the number, count the dots. (Count aloud.)

5 ●●●●	3 ●●●●●	7 ●●●

Drawing Shapes

1. Join the dots with lines to draw a shape. What shape do you get?

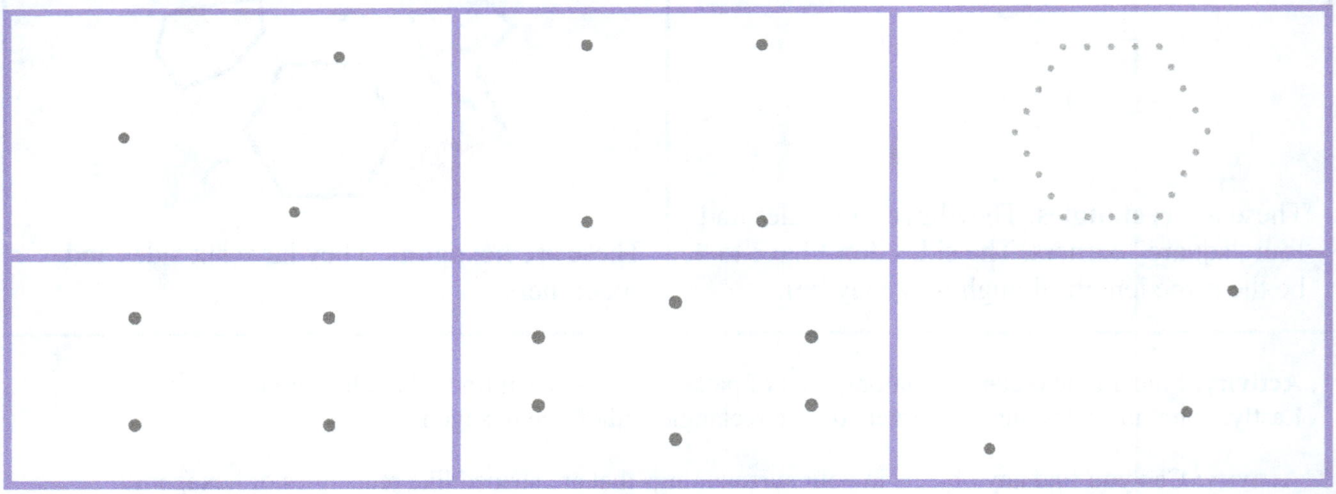

2. Draw three dots anywhere in this box. Then join the dots with lines to get a triangle.

3. In the grid of dots below, connect dots to draw: **a.** a triangle **b.** a rectangle **c.** a hexagon

4.

a. Draw a big yellow circle.	**b.** Draw a little orange square.

Shapes and Patterns

1. **Activity:** Using shape cutouts or shapes from a set of geometric shapes, let the student experiment with arranging shapes in different orders to form patterns. Ask the student to describe the patterns to you (tell you what shapes the patterns are made up of, the colors of the shapes, etc.).

2. Draw shapes and color them to continue the patterns.

a. 

b. 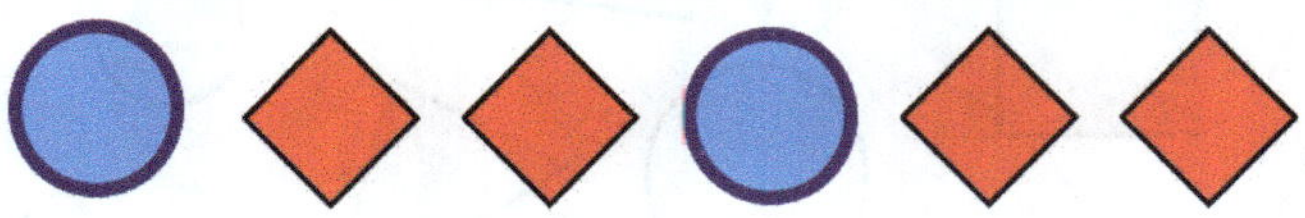

3. Continue the number pattern.

6 2 8 6 2 8 ____ ____ ____ ____ ____ ____ ____

4. Describe each shape in the pattern below to your teacher. Then continue the pattern.

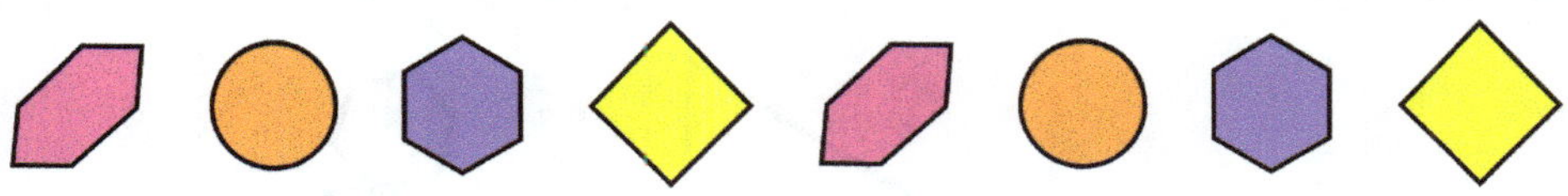

5. Subtract.

a.	b.	c.
$4 - 1 =$ ____	$9 - 1 =$ ____	$2 - 1 =$ ____
$4 - 2 =$ ____	$9 - 2 =$ ____	$2 - 2 =$ ____

6. Andy tried to continue the pattern in exercise 1b, but he made some mistakes. Cross out the shapes that don't belong in the pattern.

Puzzle Corner Draw the missing shapes in the empty spots, according to the pattern.

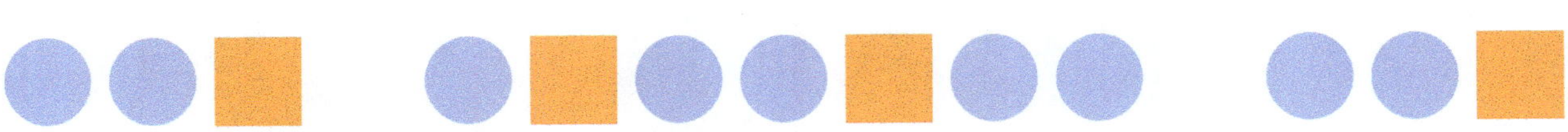

Sorting Shapes

Activity: Place an assortment of shape cutouts or pattern blocks on a flat surface (e.g. table or floor). Have the child sort the shapes and put them in separate places (such as in piles, plastic cups, etc.). Ask the child to sort the shapes in different ways, such as by color, by type of shape, and by the number of sides. After the shapes are sorted, have the child count each group of shapes and place the groups in order, from the greatest amount to the smallest amount.

1. Count each **kind of shape**. Then list how many of each shape there is, sorting the counts from the lowest to the highest.

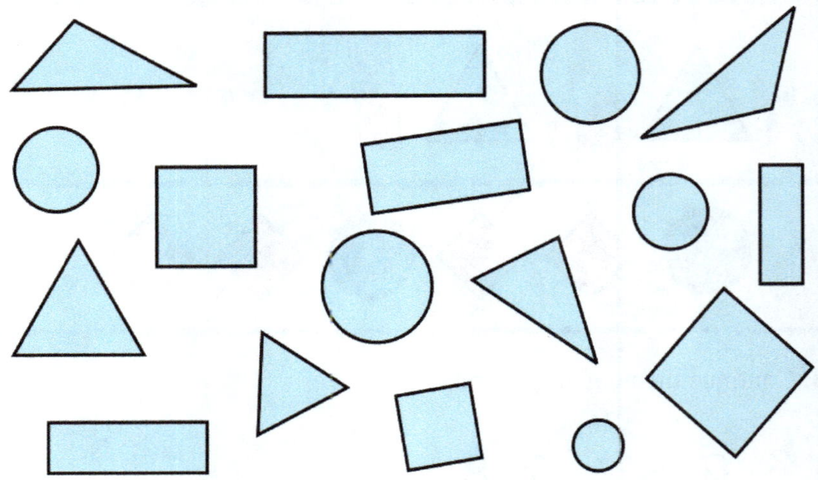

_____ _______________

_____ _______________

_____ _______________

_____ _______________

2. Color the picture following the color key.

hexagons = blue

triangles = yellow

circles = red

rectangles = orange

squares = pink

3. Sketch pictures to illustrate the situations. (Approximate sketches are sufficient.)

a. The circle is smaller than the square.	**b.** The hexagon is bigger than the triangle.

Mixed Practice 1

1. Find the missing number so that the total is 10. You can draw circles to help you.

a.  $3 + \_\_\_\_ = 10$	**b.**  $6 + \_\_\_\_ = 10$	**c.**  $4 + \_\_\_\_ = 10$

2. Count each group of dots and write the total on the line. Then circle the number that is **greater**.

a. 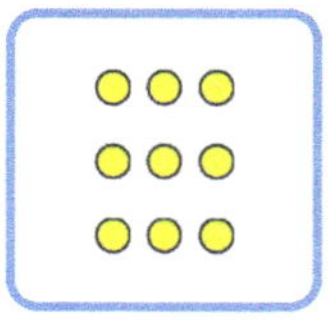 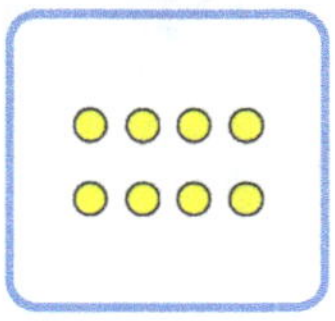 ____ ____	**b.** 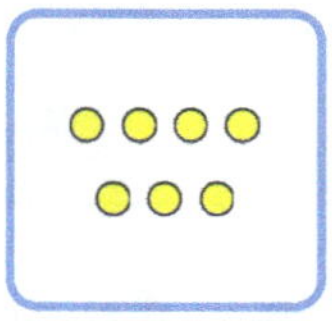 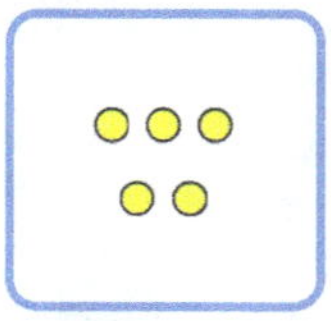 ____ ____	**c.** 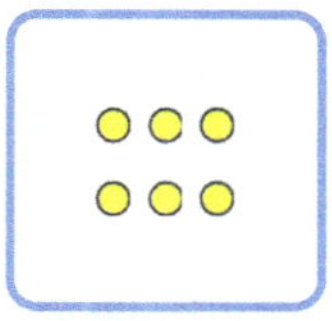 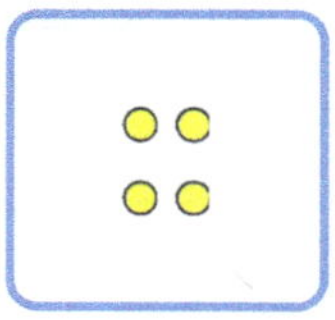 ____ ____

3. Color the picture using the key. **Note:** The student can use objects, drawings, an abacus, *etc.* to help solve the problems.

0 = black	3 = dark purple	6 = blue	9 = dark red
1 = white	4 = bright purple	7 = dark pink	10 = light purple
2 = green	5 = yellow	8 = bright red	

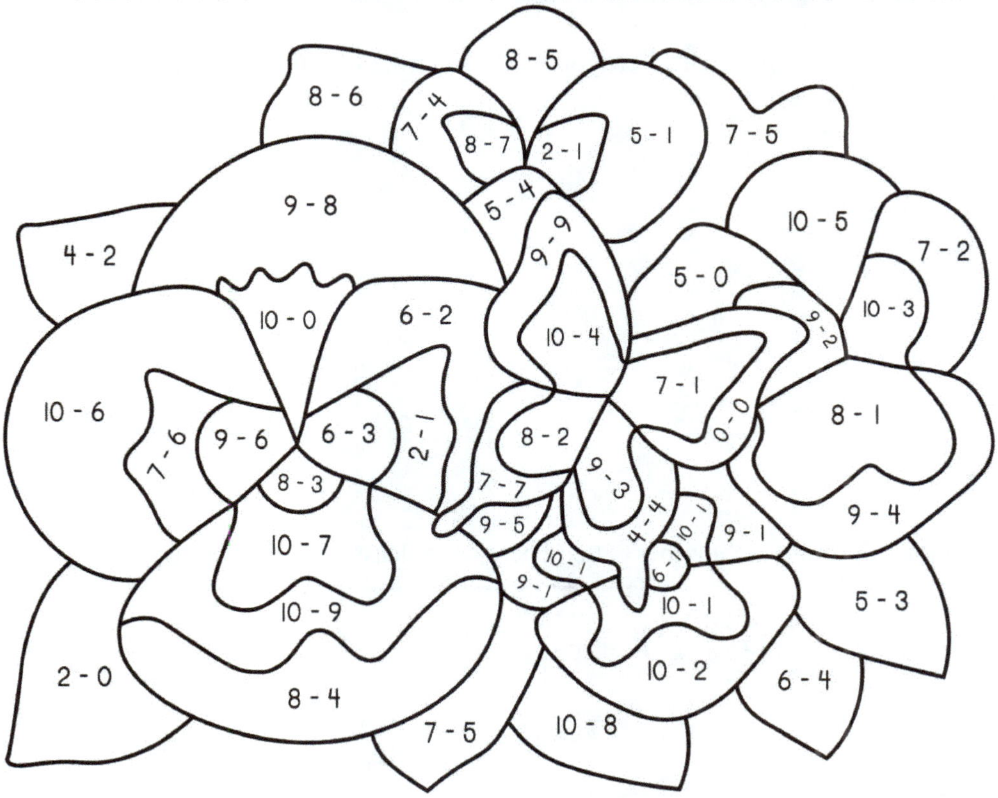

Measuring Review

1. (optional) In your house or classroom, measure how long some small items are, using toothpicks, and write their lengths. (Notes: The items should be spanned by a whole number of toothpicks, with no gaps or overlaps. You can help the child by writing the names of the objects being measured, and have the child write the numbers.)

____ _______ toothpicks

____ _______ toothpicks

____ _______ toothpicks

____ _______ toothpicks

2. Circle the object that is shorter. If the two objects are of equal height or length, circle both.

3. Draw pictures to illustrate the situations.

a. There are three crayons, and the red crayon is the longest.	**b.** Eric is taller than Lisa.

4. Illustrate the subtraction using the number path. (There are at least 2 ways to do so.)

1	2	3	4	5	6	7	8	9	10

$7 - 3 =$ ______

Weight

When we talk about how **heavy** or **light** something is, we are talking about its **weight**. Light things are easy to lift. Heavy things are not.

Look at the two animals.

Which one do you think is heavy?

Which one do you think is light?

1. **Activity:** Gather together a variety of objects, some that are easy for the student to lift and some that are harder. Possible objects include toys, cans of food, a pillow, a chair, a backpack with books in it, *etc*.

   a. Ask the child to try lift one heavy object and explain it is heavy. Do the same for one light object. Then ask the child to describe the weight of some other object. For example: "Based on what we've talked about, how would you describe *this*?" (The answer could be: "It is light.")

   Additionally, have the student give examples of objects that would be too heavy for them to lift.

   b. Have the student sort the objects into a "heavy" pile and a "light" pile.

2. Circle all of the **heavy** objects below.

3. Circle all of the **light** objects below.

Heavier and Lighter

Look at the two objects below. Which one do you think is **heavier**? Which one is **lighter**? (If you have both on hand, check and see by picking them up!)

If you said that the apple is heavier and the ruler is lighter, you're right!

1. **Activity.** Gather together objects of varying weights.

   a. Give two objects to the child and have them hold or lift them in each hand. Ask: "Which one is heavier?" or "Which one is lighter?" You can also select items that are different, yet weigh the same, such as two different kinds of canned food, and see if the student notices they are of equal weight.

   b. Choose an object. Ask the student to find an object that is heavier (or lighter) than the chosen object.

   c. Set three objects beside each other and ask the child to find the heaviest or the lightest among them.

2. Circle the heavier object.

3. Circle the lighter object.

4. Number the objects in order from <u>lightest to heaviest</u> (1, 2, 3).

Times of Day

The moon and stars are shining. It is **night**.

The sun is rising.
It is **morning**.

The sun is high in the sky.
It is **afternoon**.

1. **Activity:** Discuss the student's daily routine, asking questions such as "What do you do in the morning?", "What do you do in the afternoon?", "What do you do in the evening?", "What do you do at night?", followed by questions such as "Do you eat breakfast in the morning or in the afternoon?"

2. Match each activity with the time of day when it would normally be done. Note: There may be more than one possible correct answer for some activities, depending on the child's daily schedule.

eat breakfast

afternoon

do homework

look at the stars

evening/night

wake up

eat a picnic lunch

morning

take a bath

3. Do you normally do these activities at evening/night, in the morning, or in the afternoon? Tell your teacher.

read a book

eat supper

go to school / begin class

brush your teeth

More Practice with Measurable Attributes

1. **Activity:** Gather an assortment of objects and practice comparing their weights. Have the child hold two different objects, one in each hand. Ask, "Which one is heavier?" or "Which one is lighter?" After the child answers, repeat the exercise with different objects. Include some pairs of objects where the bigger object is lighter than the smaller object, to help show that bigger doesn't always mean heavier. For example, the child could compare a roll of paper towels with a can of beans, a balloon with an apple, etc.

2. Match each object to **all** the words that accurately describe it.
   (**Note:** Describing an object as "tall" or "short", "heavy" or "light", etc. can be very subjective, but that is fine in this exercise. The purpose here is for the child to <u>use more than one word</u> to describe the attributes of an object, not to absolutely correctly classify something as "tall", "short", "light", etc.)

heavy

small

long

tall

light

short

big

3. Circle the lighter object.

4. **Activity:** After completing exercise #2, gather an assortment of objects and have the child use multiple words to describe them. For example, a pencil might be described as short, small, and light. Prompt the child if necessary by writing descriptive words (big, small, heavy, etc.) on index cards and reading them off for the child to choose from.

5. Use two words (two adjectives) to describe this piece of yarn.

Times of Day 2

1. Do you normally do these images at evening/night, in the morning, or in the afternoon? Tell your teacher.

take a shower

eat a snack

go to the beach

get dressed

2. Listen while your teacher reads the following sentences. Cross out any that don't make sense.

a. I eat lunch in the evening. **b.** I go to bed at night. **c.** I look at the stars in the afternoon.

3. Match each activity with the time of day when it would normally be done. Note: There may be more than one possible correct answer for some images, depending on the child's daily schedule.

ride a bike

afternoon

do laundry

catch fireflies

evening/night

play basketball

wash breakfast dishes

morning

watch fireworks

4. Subtract.

a. $\begin{array}{r} 1 \\ -\ 0 \\ \hline \end{array}$ **b.** $\begin{array}{r} 5 \\ -\ 5 \\ \hline \end{array}$ **c.** $\begin{array}{r} 4 \\ -\ 3 \\ \hline \end{array}$ **d.** $\begin{array}{r} 3 \\ -\ 1 \\ \hline \end{array}$ **e.** $\begin{array}{r} 5 \\ -\ 4 \\ \hline \end{array}$

Mixed Review Chapter 5

1. Write the subtraction sentence illustrated by the arrows.

_____ − _____ = _____

2. Circle the number that is **less**. If the two numbers are **equal**, circle both numbers.

a. 7 3 **b.** 8 9 **c.** 4 4 **d.** 6 2

3. How many lemons and strawberries are there in total? Write an addition.

_____ + _____ = _____

4. What kind of shapes make up the pattern below? Draw more shapes to continue the pattern.

5. Draw a picture to illustrate the problem. Solve.

Sarah had eight carrots. Her pet bunny ate three. How many carrots does Sarah have left?

6. Add. **Note:** The student can use objects, drawings, an abacus, *etc.* to help solve these problems.

a. $9 + 1$ **b.** $0 + 6$ **c.** $4 + 4$ **d.** $5 + 2$ **e.** $2 + 3$

Chapter 5 Review

1. Draw a line to match each object with its basic shape.

2. Circle all the heavy objects in blue and all the light objects in red.

3. Draw pictures to illustrate the situations.

a. The triangle is bigger than the square.	**b.** The hexagon is smaller than the rectangle.

Chapter 6: Two-Digit Numbers & Money

Introduction

This chapter focuses on two-digit numbers, and especially on the counting sequence from 1 to 100. We will be using a 100-bead abacus (school abacus) and a visual model of sticks bundled in tens. You're welcome to also use linking cubes (Unifix cubes) and/or base ten blocks to illustrate the numbers.

We start out with numbers from 11 through 20. The child is expected to learn to write these numbers with numerals, and to decompose them as 10 plus some other number. This prepares them to learn place value with tens and ones in first grade.

For numbers from 21 through 100, the main goal is to **learn the counting sequence**. The child(ren) should learn to count from 1 to 100 by ones, and from 10 to 100 by tens — and this can happen just by rote memory. The lessons do use visuals that show the idea of tens and ones (sticks that are bundled into groups of 10), and we do practice making two-digit numbers with the abacus (linking cubes work also), but even so, the child does not have to master the concept of place value with tens and ones in kindergarten. They do not even have to learn to write the numbers from 21 to 100 with numerals. First grade is the time when we really delve into these topics. Right now, the idea is learn to count to 100, while also starting to learn about tens and ones.

The lesson "Counting Groups" is optional, because it practices counting large collections of objects, which is technically a first grade topic. The natural idea is to count these collections by first making groups of ten, but don't let the child know that at first. Let them experiment and even struggle a bit. Typically, if they try to count a large collection by ones, they will ultimately lose track and get confused. That is good in a sense, because once they find the easier way to count (or you tell them how) after struggling, they will remember it much better than if they were told in the beginning how to do it, and they just followed instructions.

In summary, students' work in kindergarten prepares them to learn about place value in first grade. Then in second grade, they learn about three-digit numbers, in third, about four-digit numbers, and in fourth, numbers through one million.

The lessons on coins are entirely optional. If your child or students have understood the basic concept of tens and ones, and are comfortable with two-digit numbers, you can proceed to this section. Note that this section does use two-digit numbers written as numerals (e.g. 58). Dimes and pennies give a nice application of tens and ones, but like mentioned earlier, understanding that, say, 42, is made up of 4 tens and 2 ones, is not fully required at kindergarten level.

The lessons that introduce the nickel take the level up yet a notch. So, use your judgment. Children that are doing well with the math concepts will probably enjoy studying coins, but don't feel pressured. This topic will be there in first and second grades also.

Our worksheet maker at https://www.mathmammoth.com/private/Make_extra_worksheets_K.htm provides extra worksheets for several of topics in this chapter.

Please also see the following pages for games that I recommend while studying this chapter.

Good Mathematical Practices

- Patterns are important in mathematics. Make sure children recognize the pattern (the structure) in the counting sequence. It is best if they notice it themselves; give them the opportunity to do so first, before pointing it out.

Manipulatives

Kindergartners will need to see and experience the quantities from 11 to 100 with physical objects. I recommend using a 100-bead abacus plus some means where the child will see the items bundled into tens. For the latter, you can use toothpicks that you bundle with rubber bands, marbles or rocks that you place into little bags, math linking cubes, or similar.

Below, I have listed a few Amazon links for your convenience.

An Abacus with 100 beads

Learning Resources 10-Row Counting Frame or Abacus for Kids
This abacus has every five beads in alternating color (red, white), which is nice.
https://www.amazon.com/dp/B01LX6K2CV/?tag=mathmammoth-20

Wooden Abacus for Kids with Counting Sticks and Number Cards
https://www.amazon.com/dp/B08SBWKJZJ/?tag=mathmammoth-20

Search Amazon's collection here:
https://www.amazon.com/s?k=school+abacus&tag=matmammoth-20

Linking Cubes

Learning Resources MathLink Cubes (100 pcs)
https://www.amazon.com//dp/B000URL296/?tag=mathmammoth-20

200 Pcs Math Linking Cubes
https://www.amazon.com/dp/B0GGCK8L8D/?tag=mathmammoth-20

Search Amazon's collection for math linking cubes:
https://www.amazon.com/s?k=math+linking+cubes&tag=mathmammoth-20

Pacing Suggestion for Chapter 6

The lessons are intended to take one day each. Feel free to take a day at times to just do activities and play games, without any written work. We want children to enjoy doing math, and not feel pressured.

The Lessons in Chapter 6	page	span	suggested pacing	your pacing
Numbers 11 to 15	35	*2 pages*	1 day	
Numbers 16 to 20	37	*2 pages*	1 day	
Counting up to 20	39	*2 pages*	1 day	
Teen Numbers & Equations 1	41	*1 page*	1 day	
Teen Numbers & Equations 2	42	*1 page*	1 day	
Mixed Practice 2	43	*1 page*	1 day	
Counting to 40, Part 1	44	*1 page*	1 day	
Counting to 40, Part 2	45	*1 page*	1 day	
Counting to 100, Part 1	46	*1 page*	1 day	
Counting to 100, Part 2	47	*1 page*	1 day	
Counting Groups (optional)	48	*1 page*	1 day	
Mixed Practice 3	49	*1 page*	1 day	
Counting to 100, Part 3	50	*1 page*	1 day	
Counting Practice	51	*1 page*	1 day	
More Practice with Numbers	52	*1 page*	1 day	
Pennies & Dimes 1 (optional)	53	*1 page*	1 day	
Practice with Dimes (optional)	54	*1 page*	1 day	
Pennies & Dimes 2 (optional)	55	*1 page*	1 day	
How Many Cents? (optional)	56	*1 page*	1 day	
Mixed Practice 4	57	*1 page*	1 day	

Nickels 1 (optional)	58	*1 page*	1 day
Nickels 2 (optional)	59	*1 page*	1 day
Counting Cents (optional)	60	*1 page*	1 day
Subtracting Cents (optional)	61	*1 page*	1 day
Going Shopping (optional)	62	*1 page*	1 day
Mixed Review Chapter 6	63	*1 page*	1 day
Chapter 6 Review	64	*2 pages*	1 day
TOTALS		*21 pages*	17 days
with optional content		*(31 pages)*	(27 days)

Games and Activities

Cups for Counting *(Counting up to 20 objects)*

You need: A variety of small objects and cups or bowls to place them in. Each cup should have between 11 and 20 objects (different amounts). A sheet of paper and pencil for the child or for each group of children.

Activity: If there is more than one student, have them work in pairs. One partner chooses a cup and dumps it out. The other counts the objects. Then the first partner checks the other one's work. The count is recorded on the paper as a single numeral, and also as a sum of ten and some other number.

For example, after counting 18 rocks, the child would record this as both 18 and as 10 + 8 on their paper, and draw a rock or rocks, or write "rocks" if they can write, to signify this is the count of the rocks. The child could also draw all the 18 rocks and circle 10 of them to show the count. If there is space on the paper, they can draw a line and record the count of some other cup on the same paper, otherwise they will need a new sheet of paper.

Count in a Circle *(The counting sequence to 100)*

Activity: This activity works best if you have more than two people. The players sit in a circle. One person (e.g. the teacher) starts the counting. Each person says the next number. Whenever anyone hits a multiple of ten (you can call them ten-numbers or whole tens), that person jumps up and sits back down, or claps their hands above their head, or does some other pre-decided movement that is kind of silly.

On each round, count from the starting number to 100. Let each player choose a starting number for one round.

Count in a Circle *(Count by tens to 100)*

Activity: This activity is just like above, but instead of counting by ones, you count by tens from 10 to 100. The player who says 100 does the silly movement, such as clapping their hands above their head.

Count with a Ball (*The counting sequence to 100*)

You need: A ball that is easy to throw and catch.

Activity: This activity works best if you have more than two people. The players sit in a circle. One person (e.g. the teacher) holds the ball. That player then throws the ball to someone else, while starting the count. The person receiving the ball then throws the ball to someone else, while saying the next counting number.

On each round, count from the starting number to 100. Let each player choose a starting number for one round.

Count with a Ball (*Count by tens to 100*)

Activity: This activity is just like above, but instead of counting by ones, you count by tens from 10 to 100.

Step into the Tens (*The counting sequence to 100*)

You need: Six sheets of paper on which you will write the numbers 10, 20, 30, 40, 50, and 60. One or two dice. Spread the sheets of paper with the numbers on the floor in order, with enough space in between so that a person can take 10 little steps going from one paper to the next.

Activity, variation 1: On your turn, throw the die. If the die shows 3, go to the paper with 30. If the die shows 4, go to the paper with 40, and so on. Then, take 10 little steps to go to the next number on paper, while counting aloud from the one number on the paper to the next.

Activity, variation 2: On your turn, throw two dice. Then you "travel" from the smaller number on the die times 10, to the larger number on the die, times 10. For example, if you get 2 and 6, you travel from 20 to 60. Again, count as you take the steps between the papers.

Board Game that Counts (*The counting sequence to 100*)

You need: Any board game where you advance a marker on a path. Number cards from 0 through 9. The standard deck will work if you consider 10 to be zero, and remove the face cards.

Activity, variation 1: On your turn, throw the die. If the die shows 3, go to the paper with 30. If the die shows 4, go to the paper with 40, and so on. Then, take 10 little steps to go to the next number on paper, while counting aloud from the one number on the paper to the next.

Activity, variation 2: On your turn, throw two dice. Then you "travel" from the smaller number on the die times 10, to the larger number on the die, times 10. For example, if you get 2 and 6, you travel from 20 to 60. Again, count as you take the steps between the papers.

Mammoths & Ladders

The traditional snakes & ladders game with a mammoth theme! This is a good game to practice the counting sequence. We have two versions, one goes from 1-50 and the other from 51 to 100.

https://www.mathmammoth.com/download/Mammoths-Ladders-50.pdf

https://www.mathmammoth.com/download/Mammoths-Ladders-100.pdf

Keep Your Card *(Numbers from 11 to 20)*

You need: A deck of number cards from 1 through 9. One number card with 10.

Game Play: Shuffle the deck and place it face down between the players. Place the card with 10 face up next to the deck. On your turn, draw one card from the deck. Add that number and 10, and say the sum out loud. Show your card to everyone. If you added correctly, you get to keep the card, so place it in your personal pile. If not, the card goes back to the bottom of the deck. Play until the deck is used up. The winner is the player with the most cards at the end of the game.

Variation: (challenge) Instead of having each player draw just one card from the deck, deal three cards to each player in the very beginning. On your turn, check if you can add two of your cards and get a sum that is more than 10. If yes, do so, and if your sum is correct, place the two cards into your personal pile. End your turn by drawing two cards from the deck, so that you have three cards again in your hand.

If your sum is not correct, those two cards go back to the bottom of the deck, and you draw two more from the deck. If you cannot find two cards that add up to something more than ten, you will draw one more card from the deck, and pass your turn.

Counting Money

You need: A bunch of coins (real or fake) to count.

Give the child an amount to make with the coins, such as 14 cents. Once the child does so, it is their turn to give you a money amount to make with the coins.

Start out with only pennies and dimes (they are the easiest to count). Then add the nickel. Instruct the child to count two nickels as a ten. The quarter is the most difficult of the coins and is pretty much beyond this level.

Note: You can ask the child to check your work, and then in turn, you check theirs. In the course of the activity, you can then sometimes make an intentional error, so that the child can discover it.

Shopping Game

You need: Various items to purchase at the store, paper, pen, coins, a bag or wallet to keep money in.

Make a play store that has various items to purchase. The prices need to be less than one dollar. The child may enjoy choosing prices, and/or writing price tags for them. Typically, it is best if the teacher is the storekeeper, at least in the beginning.

The child may not understand about change, so at first, the idea is that the child will shop in the store, and pay with exact change. With time you might be able to introduce the idea of paying with a larger coin than what the item costs, and receiving change as the difference between what the item costs and what the customer pays.

Some children may enjoy it if the storekeeper writes a receipt for every purchase. After a while, you might switch roles and let the child be a storekeeper (and possibly write receipts). All of my children enjoyed this activity very much.

Math Mammoth Practice Zone

Place Value Practice
This is an advanced activity for kindergarten. Type the two-digit number when it is given as a sum of its parts (e.g. you're shown 6 + 50, and you type 56).
https://www.mathmammoth.com/practice/place-value#questions=10&max-digits=2&mode=2

Give the Cat a Mouse: a 100-Chart Game
This is another advanced activity or game for kindergarten. You are asked to find a specific number on a 100-chart.
https://www.mathmammoth.com/practice/hundred-chart

Count Money
Practice counting coins and bills in this online game! You can choose the exact coins and bills to use, the maximum for the total amount, the maximum number of coins/bills, and more.
https://www.mathmammoth.com/practice/count-money

Further Resources on the Internet

These resources match the topics in this chapter, and offer online practice, online games (occasionally, printable games), and interactive illustrations of math concepts. We heartily recommend you take a look. Many people love using these resources to supplement the bookwork, to illustrate a concept better, and for some fun. Enjoy!

https://links.mathmammoth.com/Kch6

Scan me

Numbers 11 to 15

How many circles are there?
Write the amount on the line.

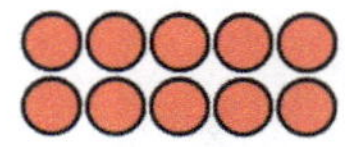

_______ Let's go on!

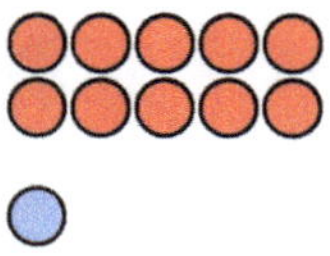

11 (eleven)

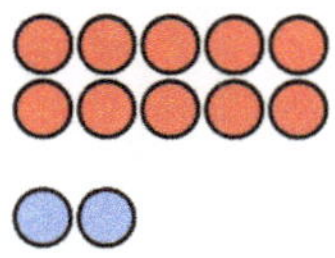

12 (twelve)

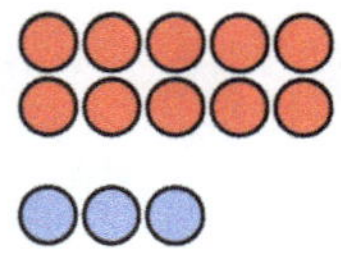

13 (thirteen)

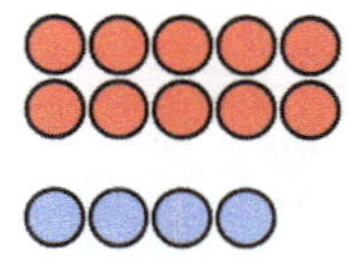

14 (fourteen)

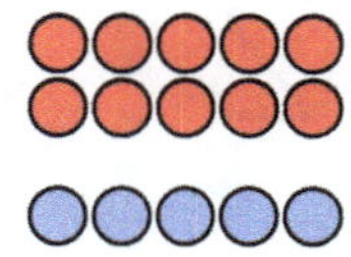

15 (fifteen)

1. **Activity:** Draw two ten-frames on paper or use a printable. You will also need 15 small objects to use as counters, such as buttons, rocks, etc. Ask the child to illustrate 10 on a ten-frame. Then, placing the two ten-frames side-by-side, count out loud with the child from 11-15 as the child places counters on the second ten-frame. Repeat as necessary. Next, have the child illustrate random numbers on the ten-frames, with special emphasis on the numbers 11-15.

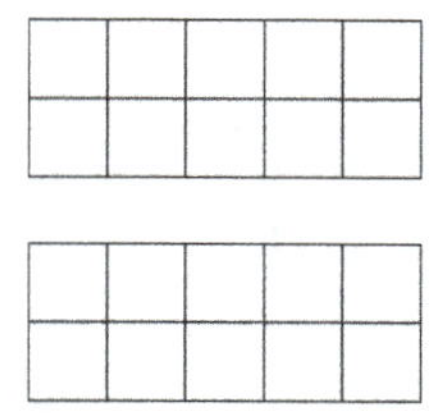

2. Circle ten objects. How many *more* objects than ten are there? Tell the total to your teacher.

a. 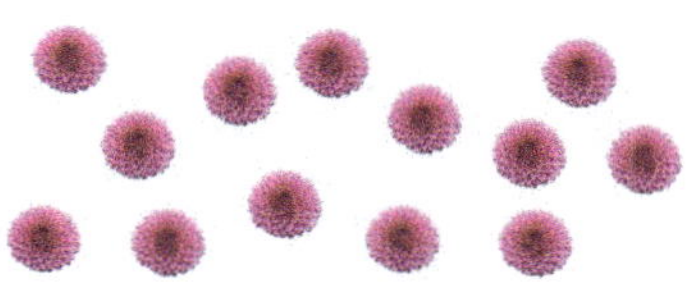

b. 

c. 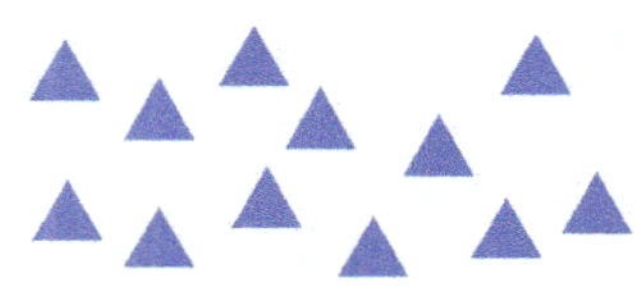

3. Draw more dots to make the target number.

a. 14 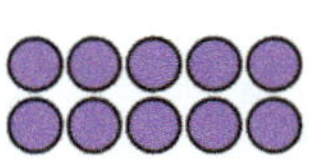

b. 11 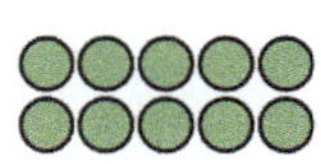

c. 13 

4. Draw a picture to illustrate the problem. Write an addition sentence and a subtraction sentence.
(*Note: there are two possible subtractions that can be written. The child can write either one.*)

Kayla bought ten cartons of juice. Some were apple juice and four were pear juice.
How many cartons of apple juice did she buy?

_______ + _______ = _______

_______ − _______ = _______

5. Practice writing the numbers on the lines.

11

12

13

14

15

6. Add.

a. 11 + 1 = _____	b. 14 + 1 = _____	c. 12 + 1 = _____

Numbers 16 to 20

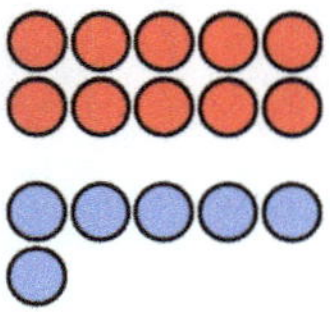 16 (sixteen)	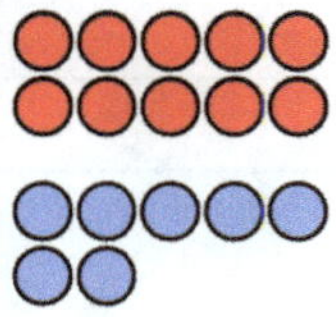 17 (seventeen)	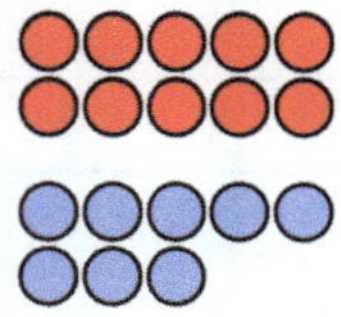 18 (eighteen)	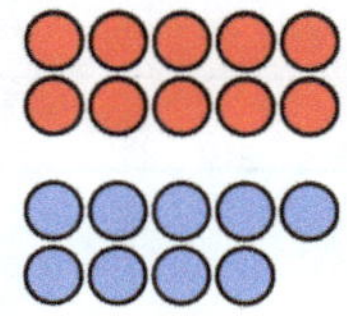 19 (nineteen)	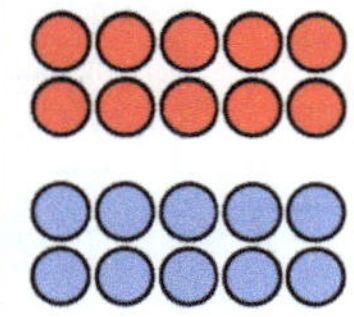 20 (twenty)

1. **Activity:** Repeat the activity from the previous lesson with a total of 20 counters.

2. First circle ten objects. Then circle some more so that the total of the two circles makes the indicated number.

a. 17 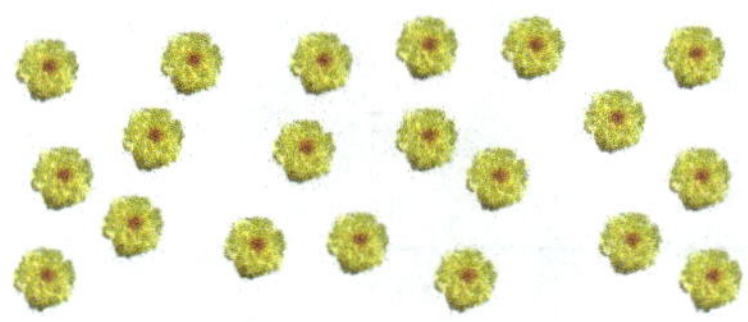	**b.** 19 
c. 18 	**d.** 16 

10	11	12	13	14	15	16	17	18	19	20

3. Draw a line to match the numbers with the groups of dots.

13	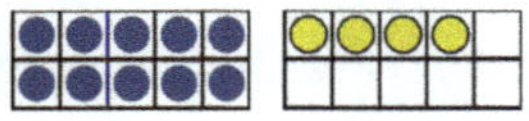
19	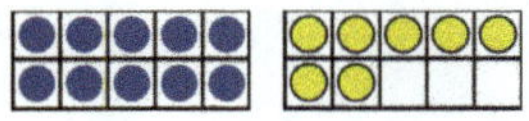
16	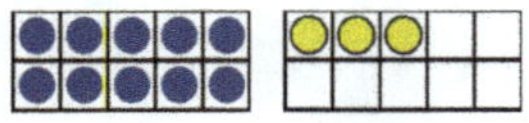
14	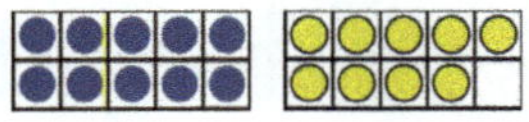
17	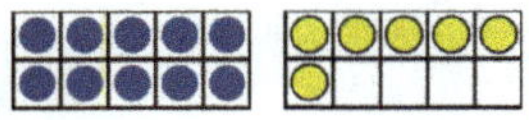

4. Add and subtract.

a. 16 – 1 = ________

b. 17 + 1 = ________

c. 20 – 1 = ________

d. 18 + 1 = ________

e. 18 – 1 = ________

f. 19 + 1 = ________

5. Practice writing the numbers on the lines.

16

17

18

19

20

Counting up to 20

1. **Activity.** First, review the counting sequence from 10 to 20 with the child(ren). Place a pile of at least 20 buttons, pencils, or some other small objects in front of the child. Ask the child (or one of the children) to choose a number between 10 and 20. Then ask them to select that number of objects from the pile, while counting each object out loud. Afterwards, return the objects to the pile, and repeat the exercise with a different amount.

2. Count the animals and circle the correct amount.

a. 

12 14 15

b. 

18 17 20

c. 

10 13 12

3. Count the shapes and write the total on the line.

a. 

____

b. 

____

4. Fill in the missing numbers.

	9	10			13			

5. Count the objects and write the total on the line.

a.  ________

b.  ________

6. Circle the indicated number of objects. How many are left? Write the total of objects that aren't circled on the line.

a. 

Circle eight objects.

How many are left?

____

b. 

Circle ten objects.

How many are left?

____

7. <u>Counting on</u> from the number, count the dots. (Start counting at the given number, and count aloud.)

13 ●●●●●	16 ●●●●	14 ●●●

Teen Numbers & Equations 1

1. **Activity.** You will need 19 items, ten of which are bundled in some manner (e.g. a rubber band around 10 toothpicks, or 10 marbles in a small plastic bag, etc.) and the rest are loose items.

   Ask the child to choose a number from 11 to 19, and then to select that number of items. (Or, choose the number yourself.) Lastly, ask the child to say a matching addition sentence (e.g. 10 + 4 = 14.)

   Conversely, choose a number of items (from 11 to 19), and ask the child how many it is.

2. Draw more sticks so that the total is the indicated number. Write the numbers in each group.

a. 16

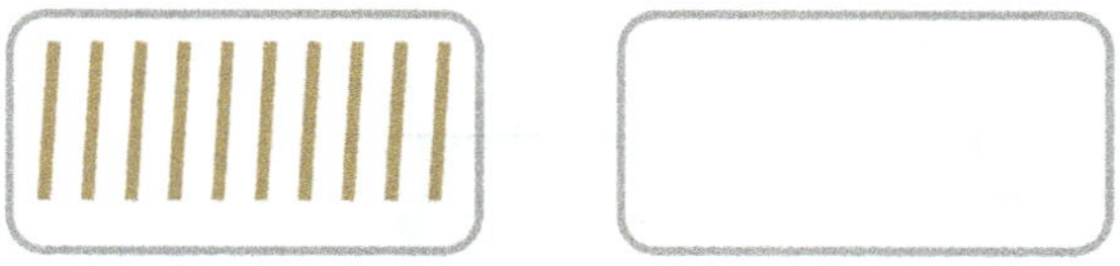

____ and ____

b. 13

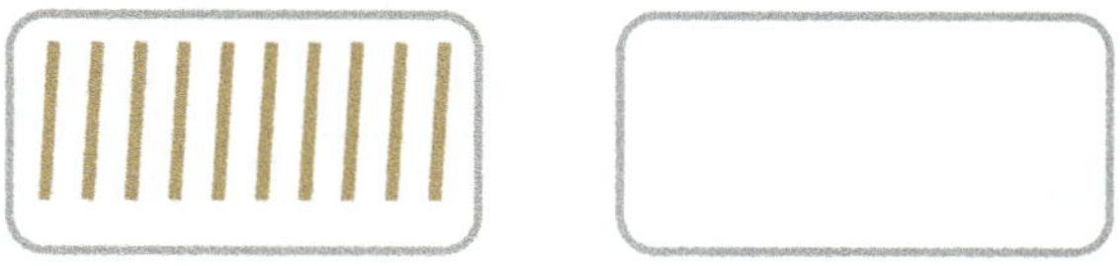

____ and ____

3. Each illustration has between 11 and 19 objects. Circle the ten objects. Then count the ones that are not in the ten-group. Lastly, write the total as a sum of "10 + some number".

a. 10 + 4

14

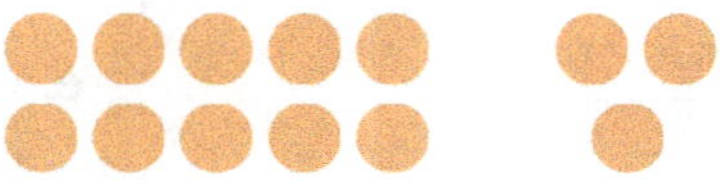

b. 10 + ____

____

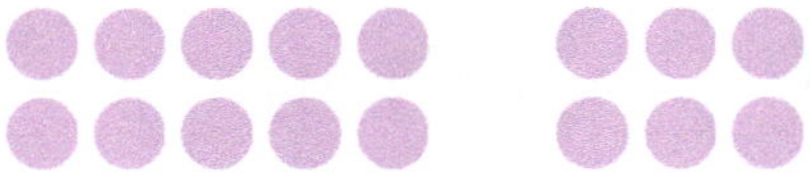

c. 10 + ____

____

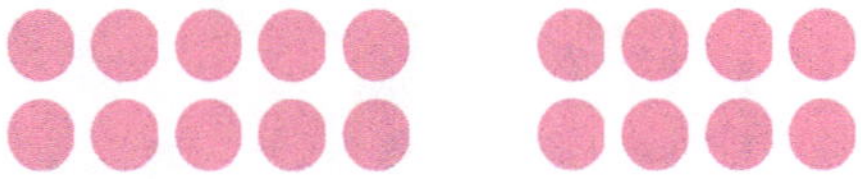

d. 10 + ____

____

4. Each group has ten circles. How many more circles do you need to complete the addition? Draw circles and write the missing number.

a. 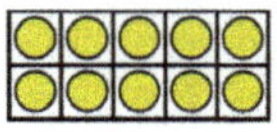

10 + ____ = 15

b. 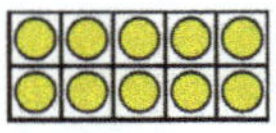

10 + ____ = 12

c. 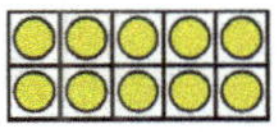

10 + ____ = 17

Teen Numbers & Equations 2

1. **Game.** Play the "Keep Your Card" game (see the chapter introduction).

2. Circle the ten objects. Then count the ones that are not in the ten-group. Next, write the total as a sum of "10 + some number". Lastly, on the bottom line, write the total.

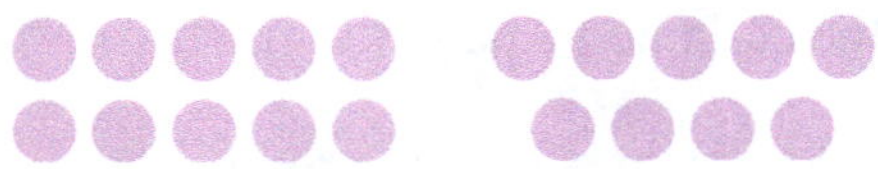

a. 10 + _____

____

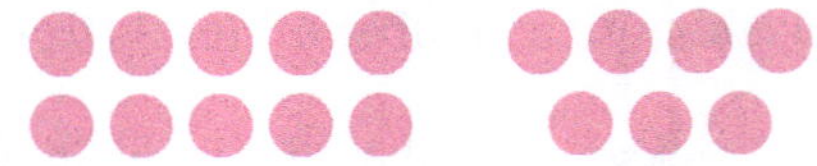

b. 10 + _____

____

3. Each group has ten circles. How many more circles do you need to complete the addition? Draw circles and write the missing number.

a. 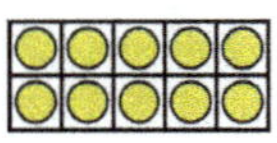

10 + _____ = 18

b. 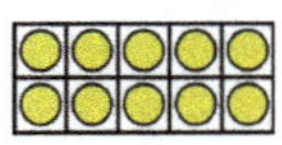

10 + _____ = 16

c. 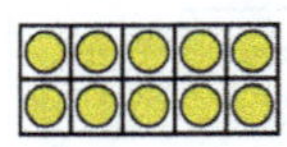

10 + _____ = 20

d. 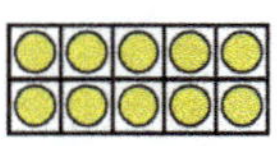

10 + _____ = 13

e. 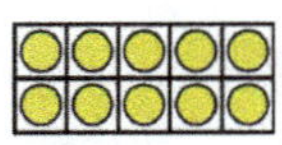

10 + _____ = 19

f. 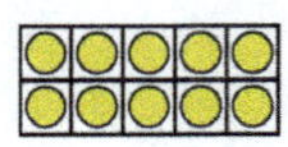

10 + _____ = 14

4. Write each number as a sum of 10 and some other number. Draw circles to illustrate the equation: draw a group of 10 circles, and some more circles.

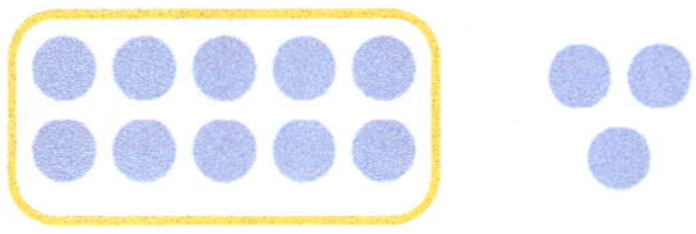

a. 13 = _______ + _____

b. 18 = _______ + _____

c. 16 = _______ + _____

d. 15 = _______ + _____

Mixed Practice 2

1. **Game.** Play an addition game (see the introduction for chapter 3 or chapter 8).

2. Add.

a.	b.	c.
$5 + 1 =$ ______	$1 + 4 =$ ______	$1 + 3 =$ ______
$2 + 2 =$ ______	$0 + 2 =$ ______	$3 + 2 =$ ______
$1 + 2 =$ ______	$2 + 3 =$ ______	$0 + 4 =$ ______

3. Subtract.

a.	b.	c.
$5 - 1 =$ ______	$4 - 1 =$ ______	$3 - 2 =$ ______
$2 - 2 =$ ______	$2 - 0 =$ ______	$5 - 3 =$ ______

4. Draw another group of circles so that there are **10 in total**. Write the numbers in each group.

a. 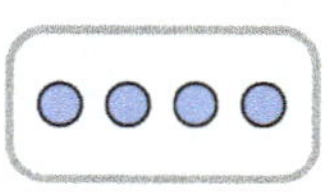

______ and ______

b. 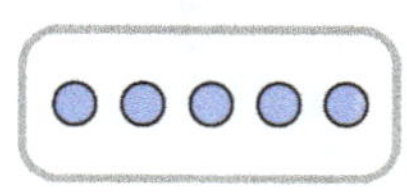

______ and ______

5. Write a **number** in the first box that is **greater** than the amount of circles in the second box.

a. ______ > 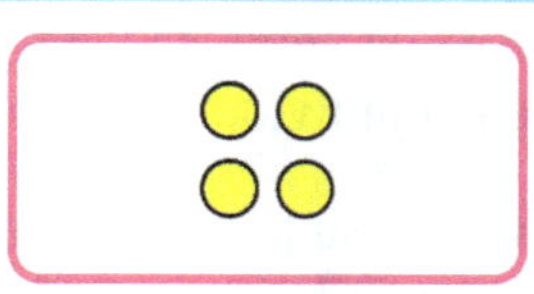

b. ______ > 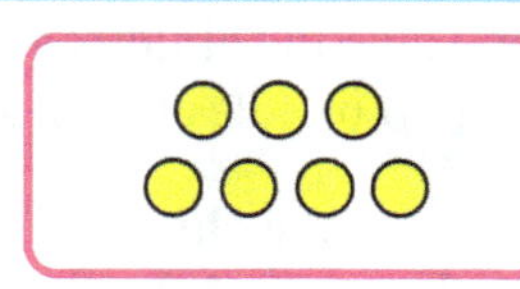

6. Draw circles in two groups, in different ways, so that the total is 7. Write how many are in each group.

a. 7	b. 7	c. 7
______ + ______	______ + ______	______ + ______

Counting to 40, Part 1

Note: You're welcome to use base ten blocks instead of sticks/toothpicks to illustrate tens and ones.

When we have many things to count, it is easier to count them if we first make bundles of 10.

Here are 10 sticks.

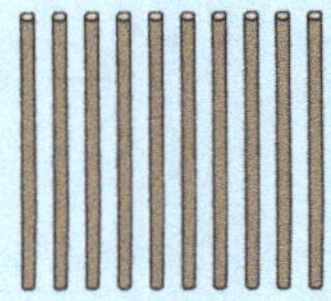

→

Let's tie the sticks into a bundle.

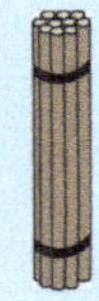

Two bundles of ten is **two tens**. We call it **twenty** and write 20.

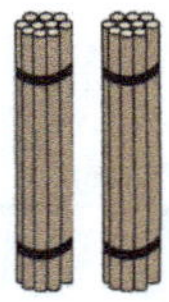

Here we have two 10-bundles and three sticks. The total is two tens and three. We call it **twenty-three**.

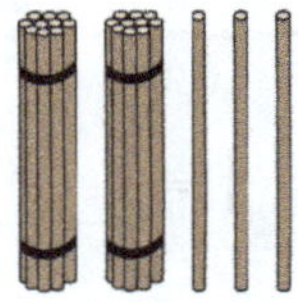

How much is this?

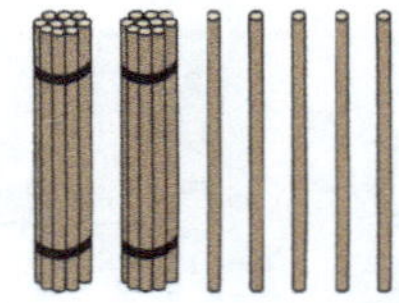

Three bundles of 10 is **three tens**. We call it **thirty** and write 30.

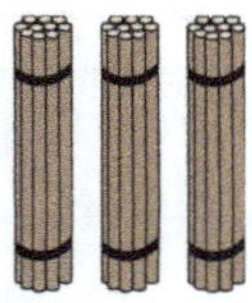

Here we have three tens and six. We call it **thirty-six**.

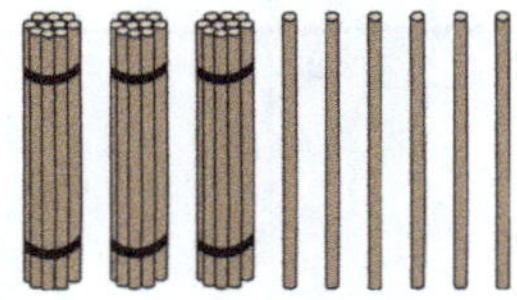

Here we have **four tens**. We call it **forty** (40).

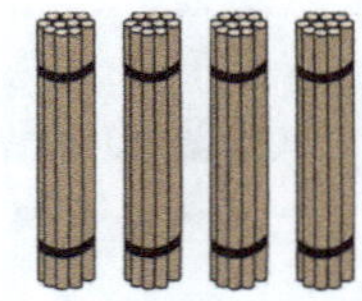

1. **Activity.** You need: Four bundles of 10 toothpicks (grouped with rubber bands), and 10 loose toothpicks.

   a. Ask the child(ren) to show you how many bundles of toothpicks make thirty, then twenty, then forty.

   b. Have the child(ren) count aloud from twenty to thirty, while you place individual toothpicks next to the two bundles of 10. After reaching thirty, exchange the ten individual toothpicks for the third ten-bundle.

   c. Have the child(ren) count aloud from thirty to forty, while you place individual toothpicks next to the three bundles of 10. After reaching forty, exchange the ten individual toothpicks for the fourth ten-bundle.

   d. Make different quantities with the toothpicks (e.g. 27, 22, 33, 38), and ask the child(ren) how many there are. Also, ask the child(ren) to form individual numbers between 20 and 40.

2. Which of these shows thirty-three? Which one shows forty? Thirty? Twenty-three?

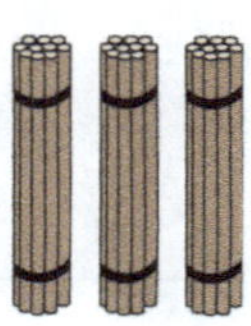

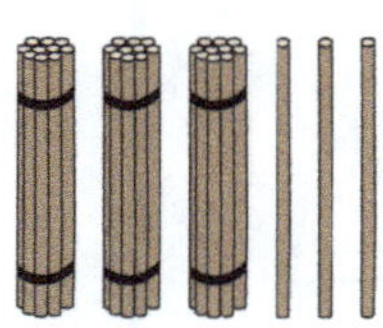

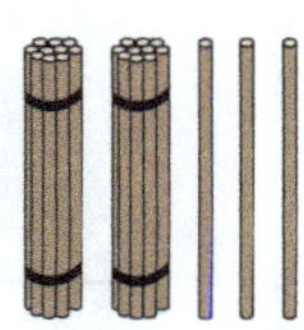

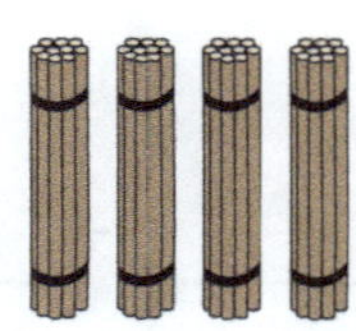

Counting to 40, Part 2

1. **Activity.** You need: a 100-bead abacus.

   a. Count with the child(ren) the beads in the first row, sliding them to the other side one by one. Continue in the same manner with the following rows, up to the last bead in the fourth row (forty), counting along.

   Once that is finished, explain that you are now going to count the beads *in groups of ten.* Sliding each group of ten beads to the other side, count together with the child(ren): ten, twenty, thirty, forty. Have the child(ren) repeat this counting by tens. You may also continue counting by tens all the way to 100.

   b. Ask the child(ren) to count from 17 to 30, then from 28 to 40. Repeat with other starting numbers, as necessary.

2. **Activity.** Work in pairs. Take turns asking your partner to make various two-digit numbers between 10 and 40 with the 100-bead abacus (e.g. 35, 22, 17, 36, 29, etc.)

3. Which number is shown on the abacus? (The beads slid to the left show the number.)

a. 

b. 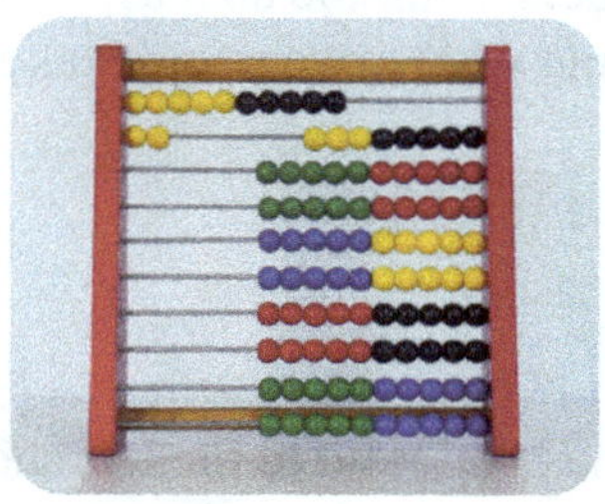

c. 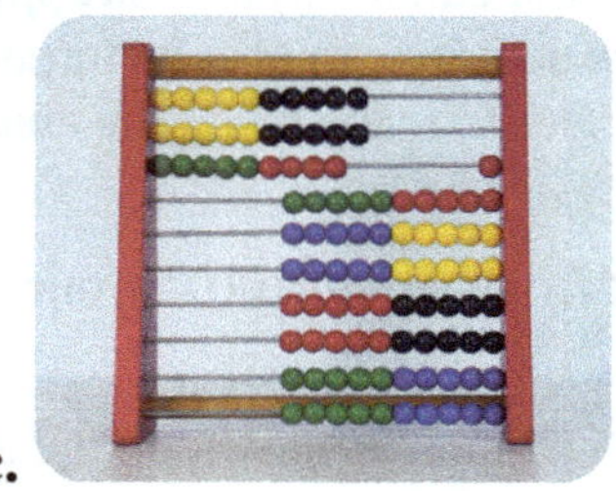

4. Match.

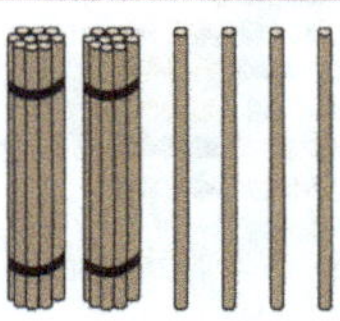
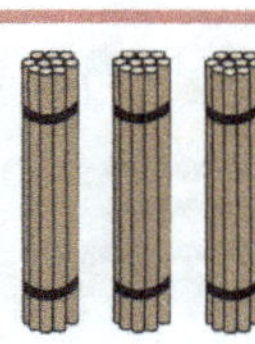
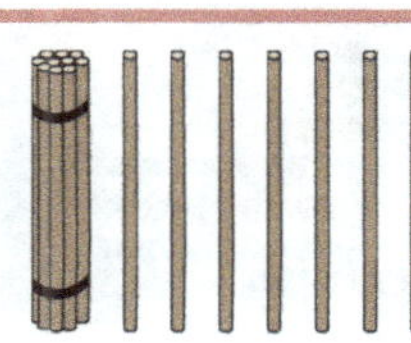

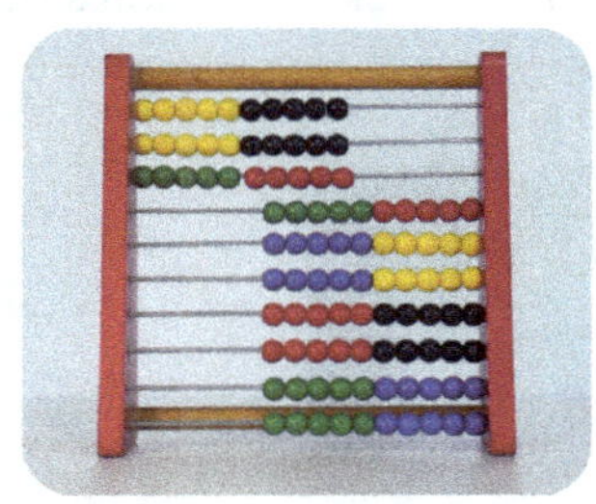

5. Draw more sticks or bundles of sticks to illustrate the given number.

a. 30 

b. 14 

c. 40 

Counting to 100, Part 1

five tens	**fifty**	50
six tens	**sixty**	60
seven tens	**seventy**	70
eight tens	**eighty**	80
nine tens	**ninety**	90
ten tens	**one hundred**	100

Count with your teacher by tens from zero to one hundred!

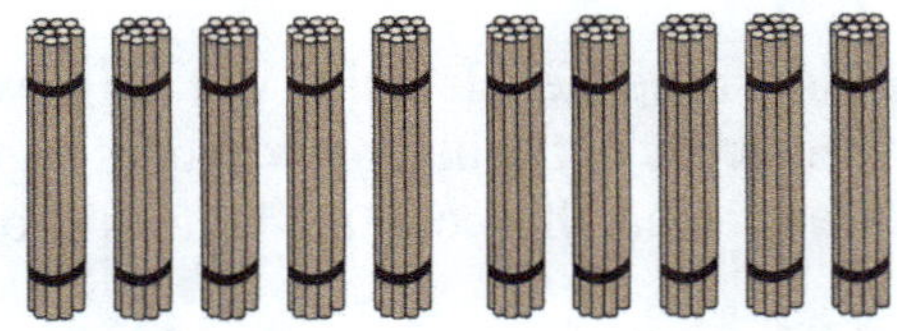

1. **Activity.** You need: a 100-bead abacus.

   a. Practice counting by tens from zero to one hundred with the child using the 100-bead abacus.

   b. Ask the child(ren) to make specific numbers using the abacus, such as twenty-four, fifty-seven, etc. This activity can also be turned into a game where two people take turns telling each other what number to make using the abacus.

2. Count from zero to 100 (by ones).

3. Which number is shown on the abacus? (The beads slid to the left show the number.)

a. 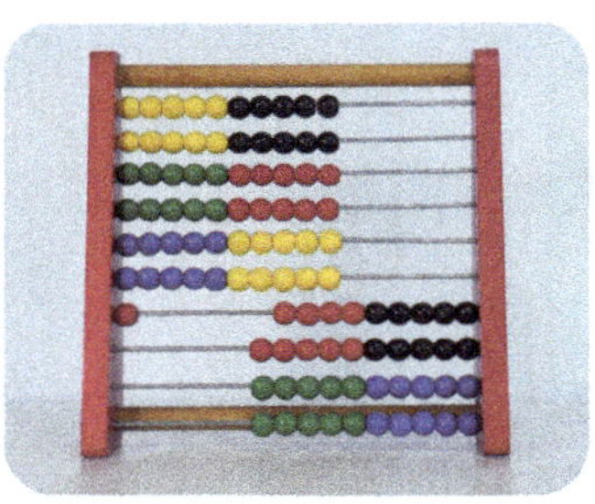

b. 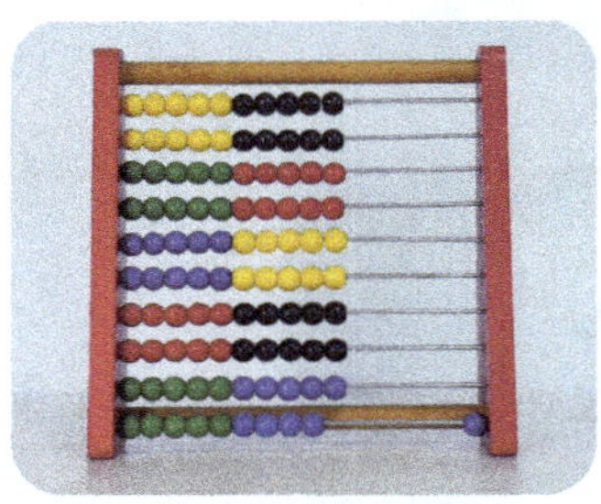

c. 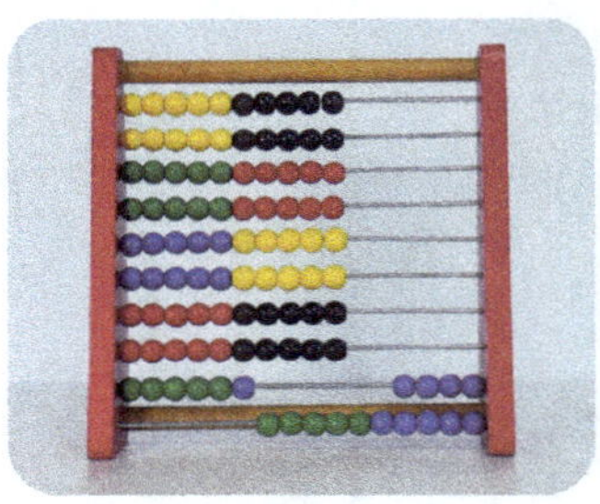

4. Draw more sticks or bundles of sticks to illustrate the given number.

a. 13 

b. 17 

c. 20 

5. How many circles are there? Write with numbers.

a. 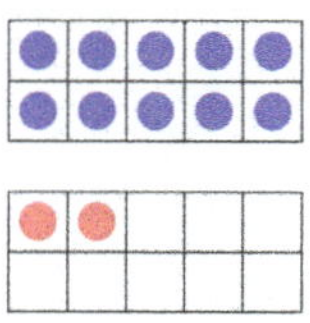 ________

b. 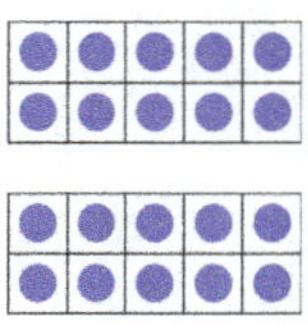 ________

c. 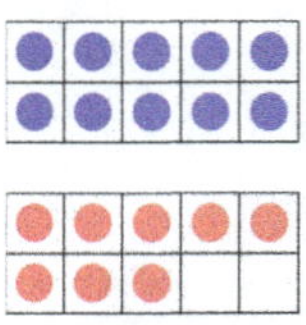 ________

Counting to 100, Part 2

1. **Activity.** You need: a 100-bead abacus.

   a. Practice counting by tens from zero to one hundred using the 100-bead abacus.

   b. Work in pairs. One person will choose a two-digit number that is 50 or more, and the other will count from that number to 100. The abacus can be used to help. Use a variety of numbers as starting points, such as: 84, 17, 35, 54.

2. Match.

90 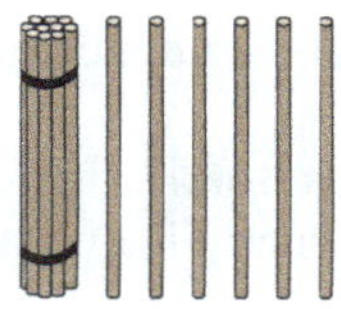

60 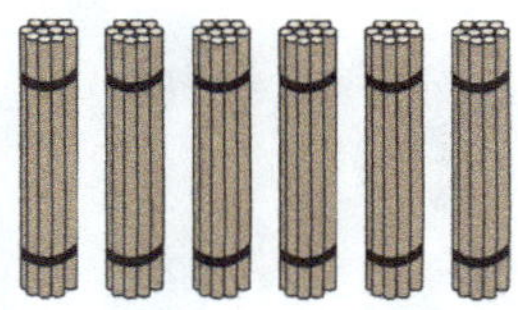

50 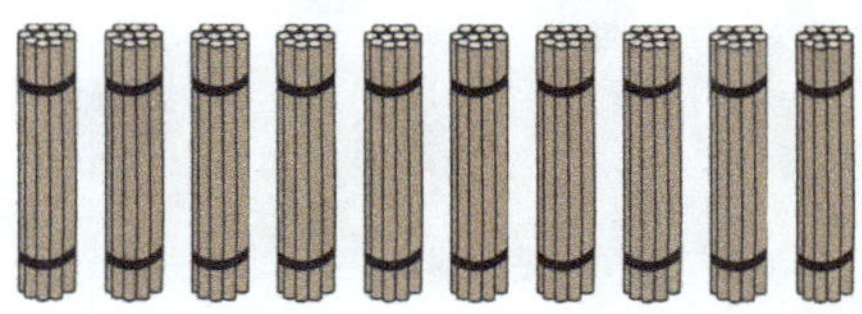

16 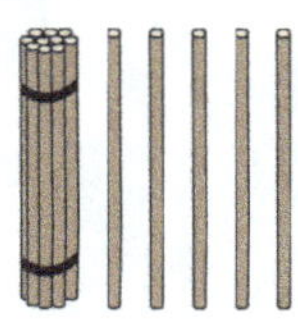

100 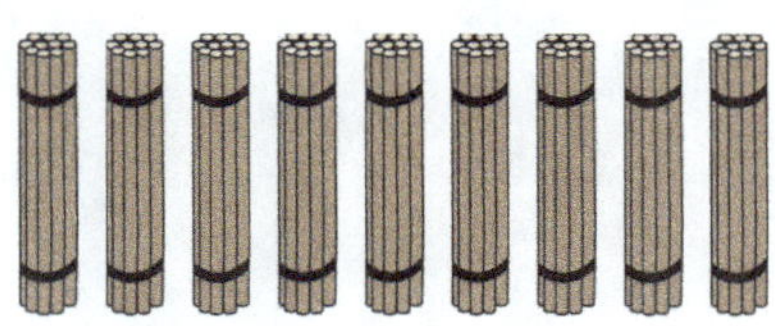

15 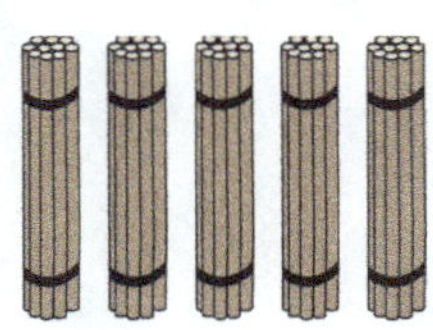

3. Write each number as a sum of 10 and some other number. Draw something to illustrate each sum.

a. 12 = ________ + ______	**b.** 19 = ________ + ______

Counting Groups

(This lesson is optional.)

1. **Counting Activity.** (challenge) You need: several different kinds of small objects to count (e.g. buttons, pennies, rocks, beans), and cups/containers to put them in. Place a different quantity of items (more than 10 but no more than 100) in each cup.

   Ask the child to count the items in each cup. If the child starts counting one by one, let them do that until they (probably) get confused and have to start over. Then ask them if they can think of a better way. Let them think first, but if they cannot find a quicker way, ultimately guide them to make groups of ten.

2. **Game**. Play a game that practices the counting sequence to 100 (see the chapter introduction).

3. (challenge) How many blueberries are there in total? How can you count the berries accurately?
   (After some struggle, you may give the child this hint: First make groups of ten.)

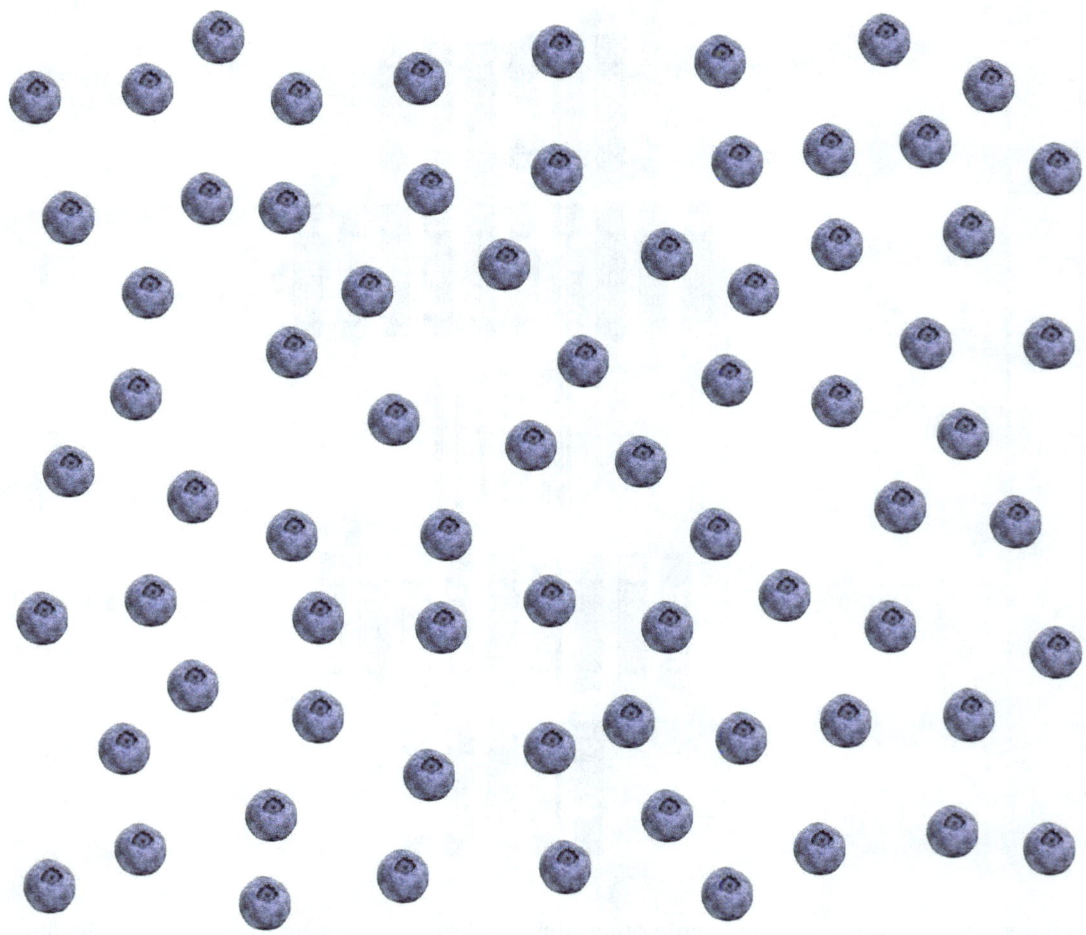

4. Write an addition sentence and a subtraction sentence about the fruits.
   (*Note: there are two possible subtractions that can be written. The child can write either one.*)

_____ + _____ = _____

_____ − _____ = _____

Mixed Practice 3

1. Circle the number that is greater (more). Practice saying these inequalities aloud.

$10 > 7$ $\quad$ $4 < 6$ $\quad$ $2 < 3$ $\quad$ $8 > 5$

2. Find the missing number so that the total is 10. You can draw dots in the ten-frames to help you.

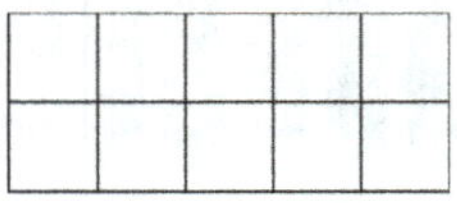 **a.** 5 + ______ = 10	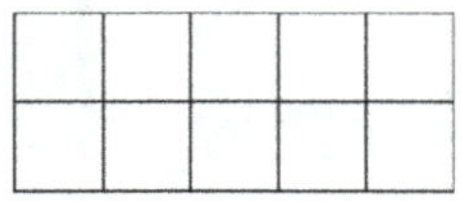 **b.** 8 + ______ = 10	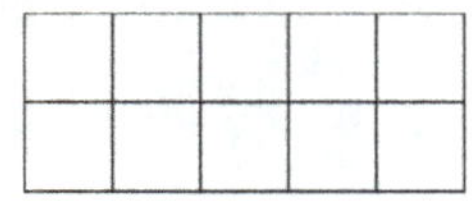 **c.** 3 + ______ = 10
d. 2 + ______ = 10	**e.** 6 + ______ = 10	**f.** 10 + ______ = 10

3. Color the circles yellow, the squares red, the triangles green, and the rectangles blue.
One shape will not be colored.

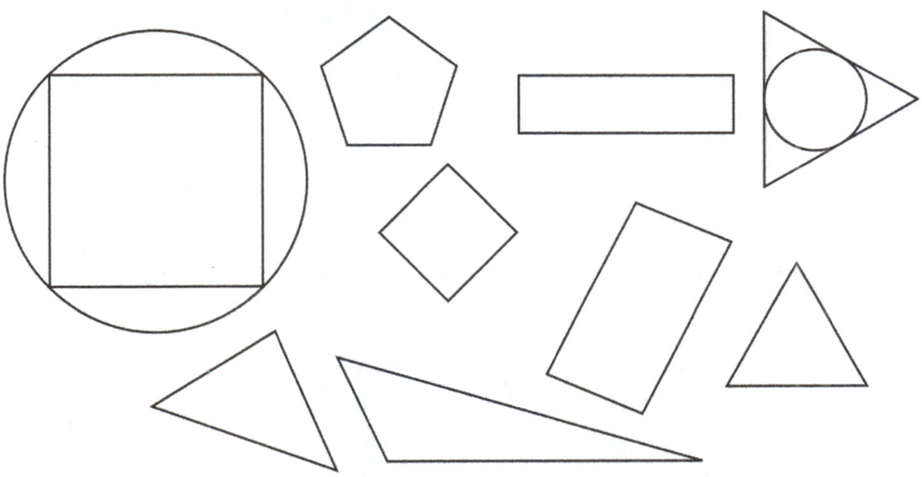

4. Which words accurately describe what is shown in each picture? Tell your teacher.

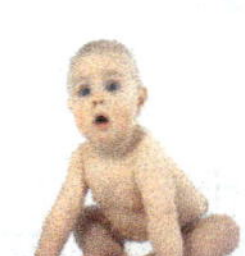 short heavy small light	 tall big light heavy
 tall long light big	 big short heavy small

Counting to 100, Part 3

1. **Game.** Play one of the games that practices the counting sequence to 100 (see the chapter introduction).

2. Choose one of the numbers represented by the sticks. Count from that number to one hundred.

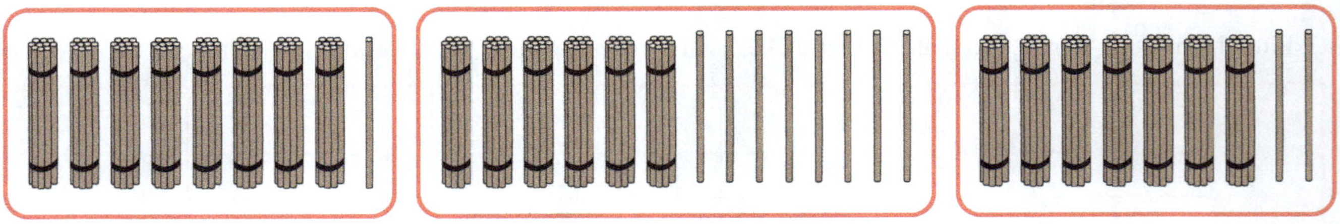

3. Draw sticks and bundles of sticks to show the given number.
(Alternatively, you can use physical manipulatives to make these numbers.)

a. fifty 50	**b.** fifteen 15
c. seventeen 17	**d.** seventy 70

4. Write the missing numbers.

_______ , 14, _______ , 16, _______ , _______ , 19 , _______

5. Write an addition.

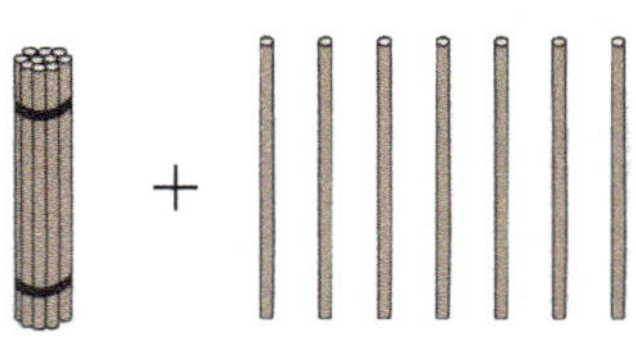	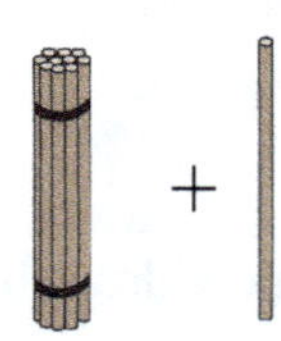
a. _______ + _____ = _______	**b.** _______ + _____ = _______

Puzzle Corner

a. 20 + 4 = _______ **b.** 60 + 20 = _______

c. 70 + 9 = _______ **d.** 42 + 10 = _______ **e.** 55 + 10 = _______

Counting Practice

1. **Game.** Play one of the games that practices counting by tens or the counting sequence to 100 (see the chapter introduction).

2. Ask the child to count from 1 to 100.

3. **Activity.** You will need: linking cubes, a 100-bead abacus, or toothpicks and toothpick bundles. Have the child make the following numbers:

twenty-seven fifty ninety-two thirty-eight seventy-seven forty-one

eighty-three sixty-five twenty-nine fifty-six ninety-four one hundred

4. How many flowers are there in total?

a.

b.

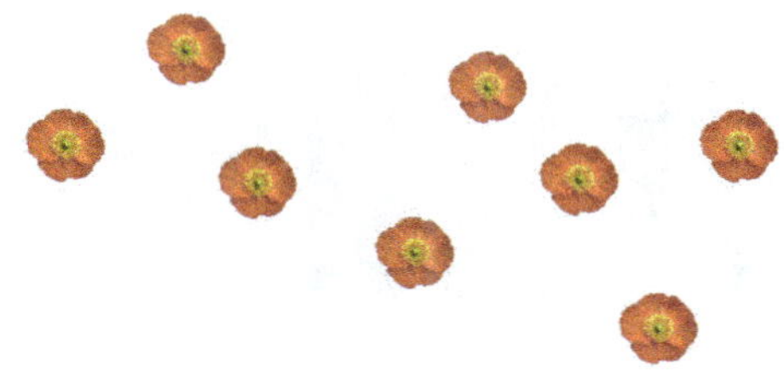

5. Match.

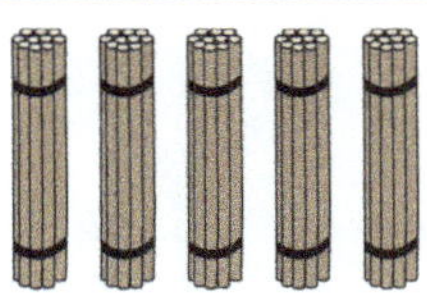

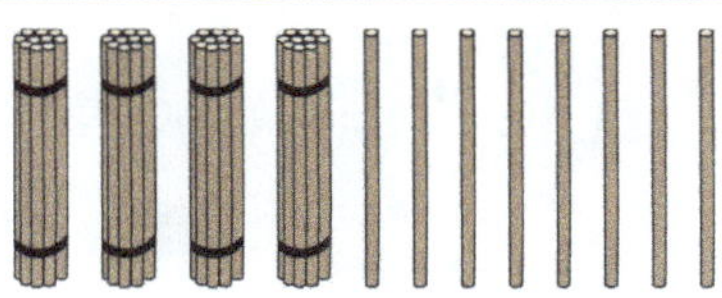

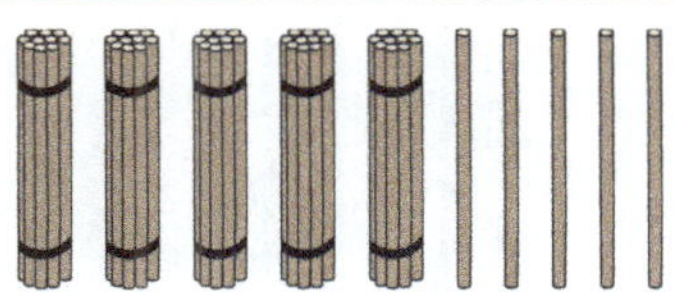

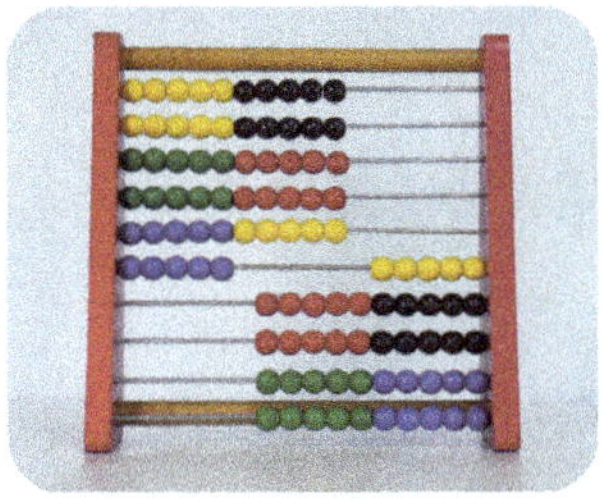

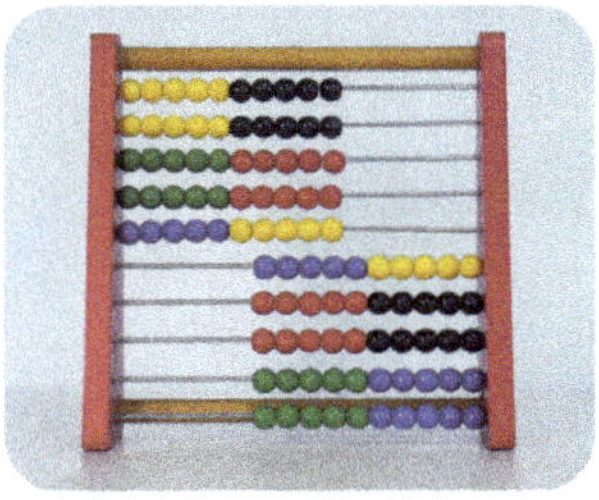

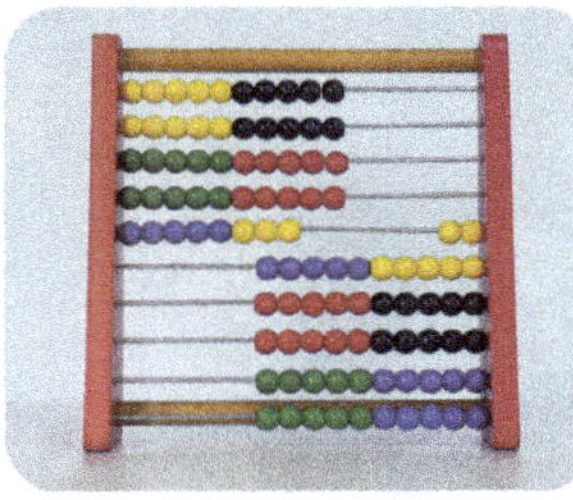

Puzzle Corner

How many flowers are there in total?

More Practice with Numbers

1. **Activity:** Using the 100-bead abacus, ask the child to make a variety of two-digit numbers (e.g. 60, 83, 29, 40, 72, 31, 57, 22, etc.). Each time, also ask the child to count forward from the number they just made until the next multiple of ten (you can call it the next ten-number or the next whole ten).

2. Ask the child to choose one of the numbers shown by the abaci, and to count from that number to 100.

   (The beads slid to the left show the number.)

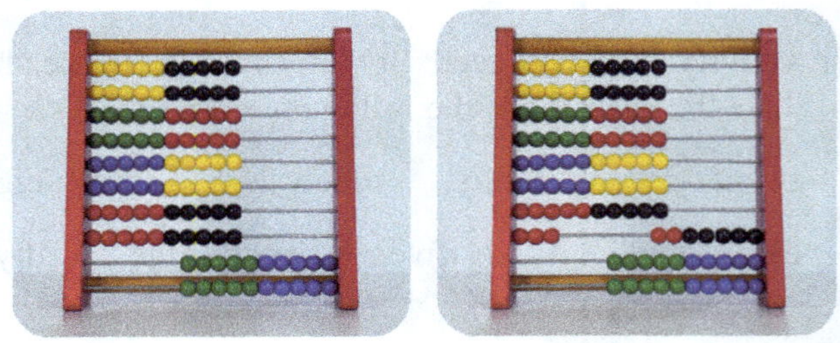

3. Draw one more ten-bundle. Now how many sticks are there? Tell your teacher.

a. 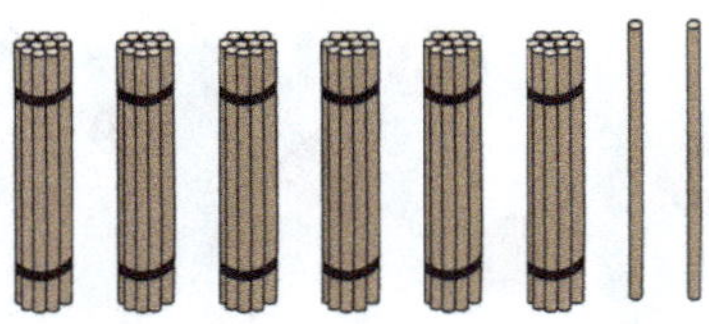

b. 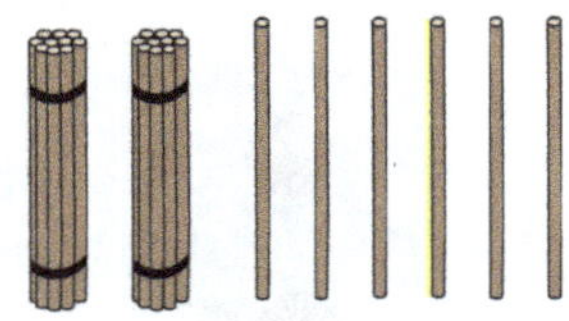

c. 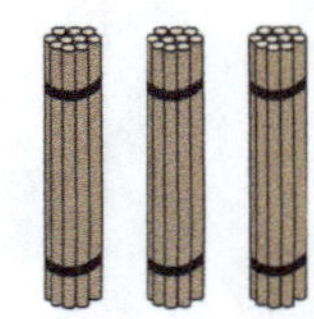

4. How many things are there in total? Write the number on the line.

a. 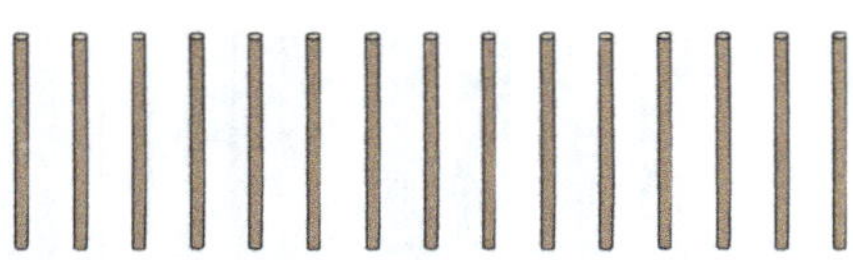 ____________

b.  ____________

5. Write an addition.

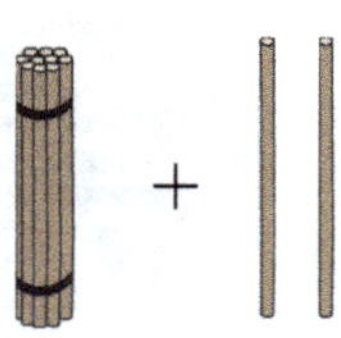

a. __________ + ______ = __________

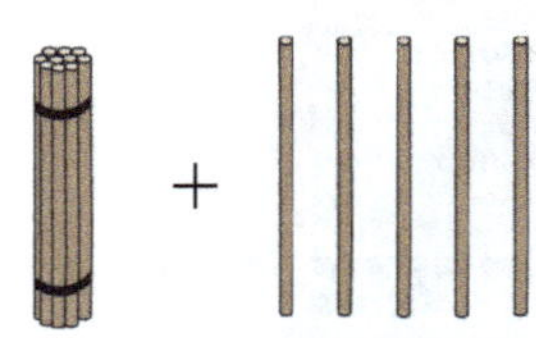

b. __________ + ______ = __________

6. Solve.

Serena saw 10 butterflies. Then she saw five more.
How many butterflies did she see in total?

Pennies & Dimes 1

front back

This coin is called a penny. It is worth one cent. We write 1¢.

front back

This coin is called a dime. It is worth ten cents (10¢).

1¢ 2¢ 3¢ 4¢

Each penny is 1 cent.
The total here is 4¢.

10¢ 11¢ 12¢ 13¢

The dime is 10 cents. Each penny is 1 cent. The total here is 13¢.

1. **Activity:** Ask the child to make specific amounts of money between 2 cents and 20 cents using pennies and dimes. You can also work in pairs, taking turns. Optionally, write down various amounts for the child(ren) to make.

2. How many cents? Circle the right number.

a. 

12¢ 16¢ 14¢

b. 

6¢ 13¢ 15¢

c. 

8¢ 17¢ 18¢

3. How many cents? Write the total on the line.

a.  _______¢

b.  _______¢

c.  _______¢

d.  _______¢

e.  _______¢

f.  _______¢

Practice with Dimes

Counting the value of dimes is essentially the same as counting by tens.

 = fifty cents, written as 50¢.

30	40	50	60	70	80	90	100
thirty	forty	fifty	sixty	seventy	eighty	ninety	one hundred

Notice how we write these with numbers!

1. **Activity:** Ask the child to make specific amounts of money with dimes only, such as 60 or 90 cents. (As a review, ask the child to also make amounts that are less than 20 cents, using one dime and pennies.) You can also work in pairs, taking turns.

2. Draw dimes (gray circles with "10" on them) or use coins to make these amounts of money.

a. 30¢	**b.** 50¢
c. 14¢	**d.** 100¢

3. Write the total amount on the line.

a.  ____________ ¢	**b.**  ____________ ¢	**c.**  ____________ ¢

4. Complete.

20¢, 30¢, __________¢ , 50¢, __________¢ , 70¢ , __________¢ , 90¢, 100¢

Pennies & Dimes 2

30¢ 31¢ 32¢ 33¢ 34¢

The three dimes make thirty cents. Each penny is 1 cent.
The total here is 34¢ (thirty-four cents).

1. **Activity:** Ask the child to make specific amounts of money between 20 and 100 cents, using dimes and pennies, such as 62 or 35 cents. You can also work in pairs, taking turns.

2. How many cents? Tell your teacher or write the total on the line.

a.  ________¢	**b.**  ________¢
c.  ________¢	**d.**  ________¢

3. Circle enough dimes to make the amount of cents.

100¢ 

60¢ 

4. Draw dimes and pennies (or use coins) to make these amounts of money.

a. 34¢ (thirty-four cents)	**b.** 45¢ (forty-five cents)	**c.** 62¢ (sixty-two cents)

How Many Cents?

1. How many cents? Tell your teacher, or write the total on the line.

a.  _________¢

b.  _________¢

c.  _________¢

d.  _________¢

2. Draw coins to make these amounts (gray circles with "10" for dimes, and orange circles with "1" for pennies).

a. 18¢

b. sixty cents

3. Match the groups of coins with the correct amounts of cents.

42¢
(forty-two cents)

24¢
(twenty-four cents)

23¢
(twenty-three cents)

4. Do the coins add up to the correct amount? If yes, circle them. If no, cross them out.

a. 16¢

b. 23¢

c. 12¢

Mixed Practice 4

1. **Activity:** make the following numbers using small objects, linking cubes, or a 100-bead abacus:

one hundred   twenty-six   eighty-eight   thirty-five   sixty-nine   ninety-three   fifty-two

2. Circle both the longest and the shortest object.

3. Draw eight dots. Then separate them into two groups, in a *different* way each time. Write how many are in each group.

a. 8	**b.** 8	**c.** 8
_____ and _____	_____ and _____	_____ and _____

4. Add and subtract. If the answer is 4, color the square **blue**.
(You can also choose colors for the other answers, as you like.)

1 + 0	2 + 0	3 + 0	4 + 0	5 + 0
1 + 1	2 + 1	3 + 1	4 + 1	5 + 1
1 + 2	2 + 2	3 + 2	4 + 2	5 + 2
1 + 3	2 + 3	3 + 3	4 + 3	5 + 3
1 + 4	2 + 4	3 + 4	4 + 4	5 + 4

Nickels 1

 This coin is a **nickel**. It is worth five cents (5¢).	 5¢ 6¢ 7¢ 8¢ The nickel is 5 cents. Each penny is 1 cent. The total here is 8¢.

1. **Activity:** Ask the child to make specific amounts of money between 5 cents and 20 cents using dimes, nickels, and pennies. Start with amounts that only require a nickel and some pennies.

2. How many cents? Write the total on the line.

a.  ______¢	**b.**  ______¢
c.  ______¢	**d.**  ______¢
e.  ______¢	**f.**  ______¢

3. Draw more coins to make these amounts of money.

a. 7¢ 	**b.** 9¢ 
c. 11¢ 	**d.** 15¢ 
e. 10¢ 	**f.** 12¢ 

Nickels 2

5¢ + 5¢ = 10¢

Two nickels make 10 cents in total.

10¢ 11¢ 12¢ 13¢ 14¢

The two nickels make 10 cents. Each penny is 1 cent. The total here is 14¢.

1. How many cents? Write the total on the line.

a.  ________¢

b.  ________¢

c.  ________¢

d.  ________¢

e.  ________¢

f.  ________¢

2. Draw nickels and pennies to make the amounts of money. Do *not* use dimes.

a. 16¢	**b.** 13¢

3. Draw only nickels to make 20 cents.

Puzzle Corner Draw nickels (only nickels!) to make sixty-five cents.

Counting Cents

1. **Activity:** Using real dimes, nickels, and pennies, make these amounts of money.

35¢ 22¢ 54¢ 29¢ 60¢ 43¢ 37¢ 56¢ 65¢

2. How many cents? Count the coins and write the total on the line.

a.  ________¢

b.  ________¢

c.  ________¢

d.  ________¢

3. Match the groups of coins with the correct amounts of cents.

25¢ 

19¢ 

33¢ 

47¢ 

4. Draw two nickels in each group. Write the *new* total on the line.

a.  ________¢

b.  ________¢

Subtracting Cents

1. **Activity (Optional)** - You need: dimes, nickels, pennies, and an assortment of small objects. Label each of the small objects with a "price tag", making the price some small amount, such as 37¢ (less than a dollar). Give the child a quantity of dimes and pennies, and ask them to make the amount of cents that is on each price tag. **Variation:** Use small amounts of some kind of snack instead of small objects, and let the child enjoy a fun treat at the end of the activity!

2. "Take away" the indicated amount of cents by crossing out coins. How many cents are left? Write the new total on the line.

a. Cross out 10¢.

_______¢

b. Cross out 5¢.

_______¢

c. Cross out 7¢.

_______¢

d. Cross out 3¢.

_______¢

3. Cross out the coins that Eric will use to buy the stamp. How much does he have left?

Eric had 84¢. Then he bought a stamp for 73¢.

Now Eric has _______¢ left.

4. Draw coins to show the *original* amount of cents. Cross out the coins that the child will use to buy the item. How much does the child have left? Alternatively, you can use physical coins.

a. Danny had 49¢.
Then he bought a banana for 26¢.

b. Melissa had 80¢. Then she bought a sticker for 30¢.

Going Shopping

1. This is how much money you have. Cross out the coins you need to buy the item.
   (Optional: How much do you have left?)

a. You have:

You buy a pencil sharpener for 77¢.

b. You have:

You buy glue for 64¢.

2. Solve.

a. You have this much money:

You buy a piece of candy for 5¢.
How much money do you have left?

b. Jessica has 45¢. She buys a toy at a yard sale for 10¢. How much money does she have left?
(Hint: draw Jessica's coins.)

3. Circle the subtraction sentence that matches the picture.

a. 

19¢ − 5¢ = 14¢ OR 18¢ − 5¢ = 13¢

b. 

25¢ − 3¢ = 22¢ OR 20¢ − 3¢ = 17¢

4. Write a subtraction sentence to fit the picture.

a. 

________ − ________ = ________

b. 

________ − ________ = ________

Mixed Review Chapter 6

1. Number the objects in order from <u>lightest to heaviest</u> (1, 2, 3).

☐ ☐ ☐

2. Draw lines to connect the dots. Use a ruler. What kind of shape did you draw?

3. Decompose 10 into pairs in different ways. Draw two groups of fish that make 10, and write the addition that they show. (*Note: there are many possible decompositions that can be done.*)

a.

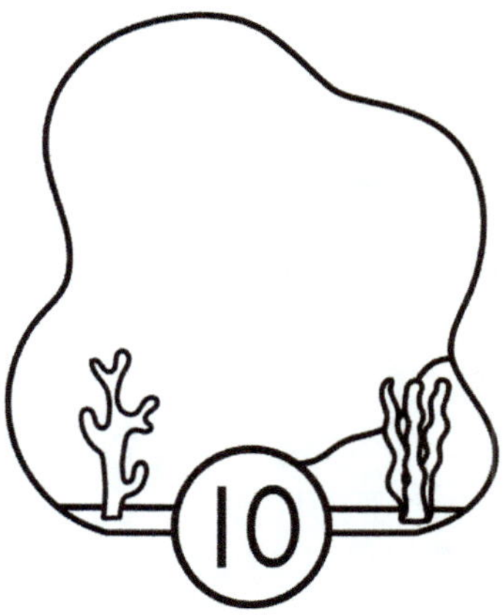

____ + ____

b.

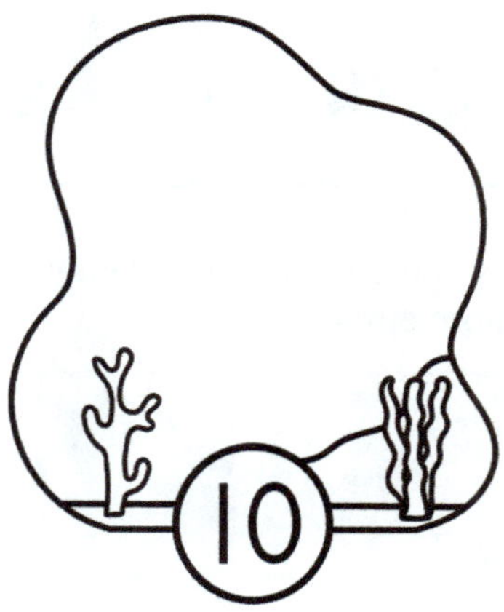

____ + ____

c.

____ + ____

4. Solve.

a. Mommy made 10 sandwiches. Daddy ate three. How many sandwiches are left?

b. Kendra picked three red flowers and four yellow flowers. How many flowers did she pick in total?

5. Add.

a. 14 + 1 = ______

b. 11 + 1 = ______

c. 19 + 1 = ______

Chapter 6 Review

1. Count the objects and write the total on the line.

a.

____

b.

____

c.

____

2. Write each number as a sum of 10 and some other number. Draw circles to illustrate the equation: draw a group of 10 circles, and some more circles.

a. 17 = ________ + ______	**b.** 14 = ________ + ______

3. Count from this number 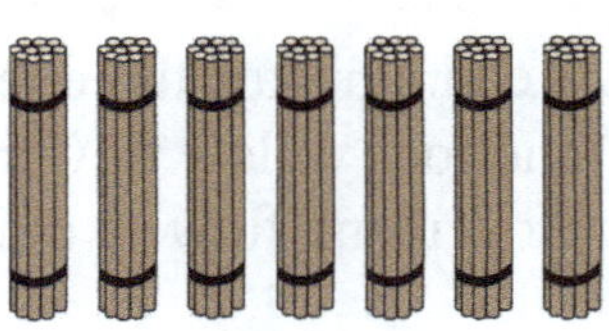 to 100.

4. Say the number word for these amounts.

a. 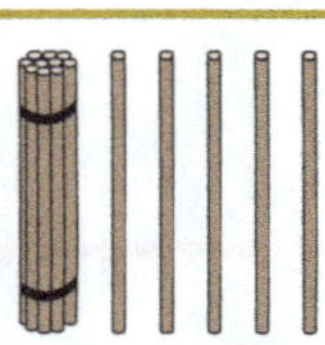	**b.** 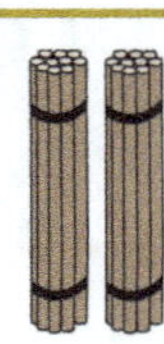	**c.** 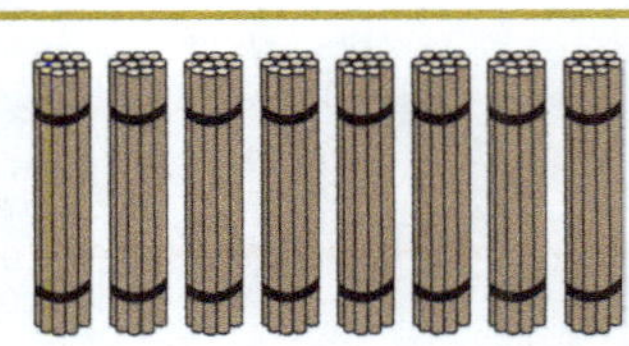
d. 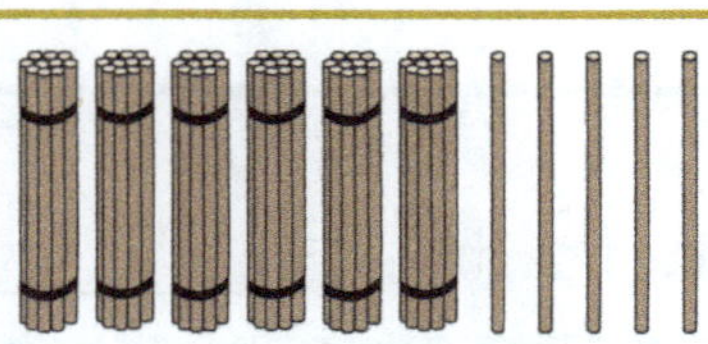	**e.** 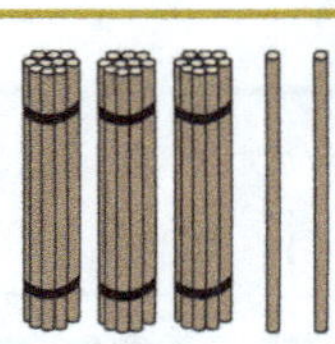	**f.** 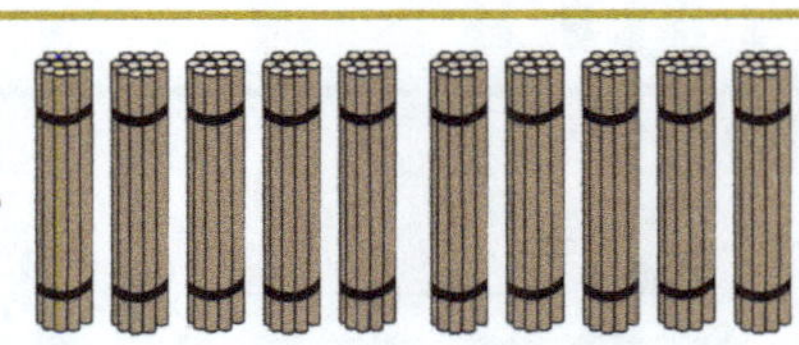

5. (challenge) How many stars are there in total?

6. (Optional) How many cents? Write the total on the line.

a.

________¢

b.

________¢

7. (Optional) You have this much money:

You buy a banana for 26¢.

How much money do you have left?

8. (Optional) Count the coins. Is the given amount correct? If yes, circle the coins. If no, cross them out.

a.

32¢

b.

28¢

c.

49¢

Chapter 7: Shapes & Position Words
Introduction

Here in the seventh chapter children study shapes some more and learn how to describe the position of an object in relation to another object.

Children learn about basic solids (3-D shapes) — specifically cubes, spheres, cylinders, and cones. They work with a set of geometric solids and locate everyday objects in the shape of these solids. They learn to identify shapes as two-dimensional (lying in a plane, "flat") or three-dimensional ("solid").

Then we also spend some time analyzing and comparing shapes. This means describing the similarities and differences of shapes, such as comparing the number of sides and vertices (corners) the shapes have, comparing their color and overall size, and even checking if a shape has equal sides or not.

Students will continue and advance in describing basic attributes of shapes in first and second grades, preparing them to reason about the shapes in a more formal manner later on. In 4th and 5th grades, they will then classify two-dimensional figures based on specific attributes related to angles and sides.

Also in this chapter, children will compose new shapes from smaller shapes. In addition to the work in the lessons, I recommend that you also get a set of geometric shapes, such as a tangram set. Look for one that comes with designs that can be built with the shapes (usually a set of cards).

One additional topic for this chapter is ordinal numbers. Most children probably learn these through everyday conversations, but it is good to check that they have indeed learned these basic words.

Again, there are lots of hands-on activities in the lessons of this chapter, and they are important. Besides the lessons, feel free to take a day or two just to play addition and subtraction games for review (see the introductions for earlier chapters).

Geometric Pattern Blocks and Tangram-like games

I recommend that you get a set of geometric 2-D shapes. This could be a set of pattern blocks or a tangram-like set of shapes. The latter usually come with design cards with composite shapes for the child to make. (The links below are the same as in the chapter 5 introduction.)

Coogam Wooden Pattern Blocks Set 130PCS
https://www.amazon.com/dp/B07MYYK64R/?tag=mathmammoth-20

Beestech Wooden Tangram game with a puzzle Book
https://www.amazon.com/dp/B07Y37NNSR/?tag=mathmammoth-20

Vanmor Wooden Tangram set with shape cards
https://www.amazon.com/dp/B07XJ23KPG/?tag=mathmammoth-20

Wooden pattern blocks shape set with pattern cards
https://www.amazon.com/dp/B0CMTB3DR2/?tag=mathmammoth-20

GEMEM wooden pattern blocks set with design cards
https://www.amazon.com/dp/B07QPS5CFL/?tag=mathmammoth-20

Pacing Suggestion for Chapter 7

The Lessons in Chapter 7	page	span	suggested pacing	your pacing
Shapes Review	69	*1 page*	1 day	
Solids	70	*1 page*	1 day	
2-D Shapes and 3-D Shapes	71	*1 page*	1 day	
Shapes and Everyday Objects	72	*1 page*	1 day	
Building 3-D Shapes	73	*1 page*	1 day	
Comparing Shapes 1	74	*1 page*	1 day	
Comparing Shapes 2	75	*1 page*	1 day	
Position Words 1	76	*1 page*	1 day	
Position Words 2	77	*1 page*	1 day	
Mixed Practice 5	78	*1 page*	1 day	
Ordinal Numbers 1	79	*1 page*	1 day	
Position Words 3	80	*1 page*	1 day	
Ordinal Numbers 2	81	*1 page*	1 day	
Comparing Shapes 3	82	*1 page*	1 day	
Drawing Shapes 2	83	*1 page*	1 day	
Shapes and Order	84	*1 page*	1 day	
Composing Shapes 1	85	*1 page*	1 day	
Composing Shapes 2	89	*1 page*	1 day	
Mixed Review Chapter 7	91	*2 pages*	2 days	
Chapter 7 Review	93	*1 page*	1 day	
TOTALS		*21 pages*	21 days	

Further Resources on the Internet

These resources match the topics in this chapter, and offer online practice, online games (occasionally, printable games), and interactive illustrations of math concepts. We heartily recommend you take a look. Many people love using these resources to supplement the bookwork, to illustrate a concept better, and for some fun. Enjoy!

https://links.mathmammoth.com/Kch7

Scan me

Shapes Review

1. Say the name of each shape out loud. How many sides does each shape have? Write it on the line or tell your teacher.

________ ________ ________ ________ ________

2. Color the shapes in the picture according to the color key.

circle = yellow

square = pink

triangle = green

rectangle = blue

hexagon = purple

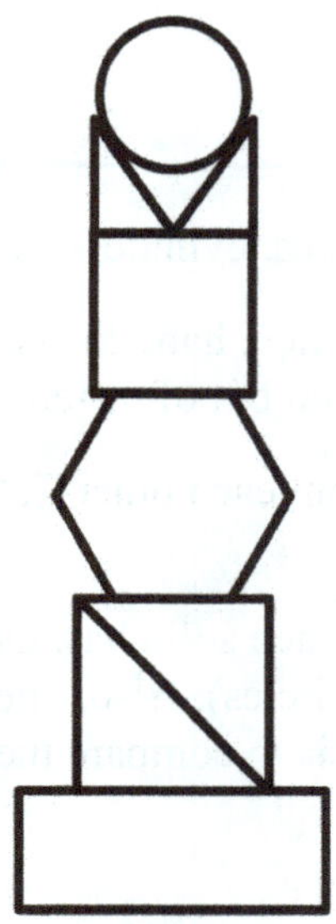

3. Draw pictures to illustrate the situations.

a. The yellow square is bigger than the red circle.

b. The pink triangle is smaller than the brown rectangle.

4. Continue the pattern.

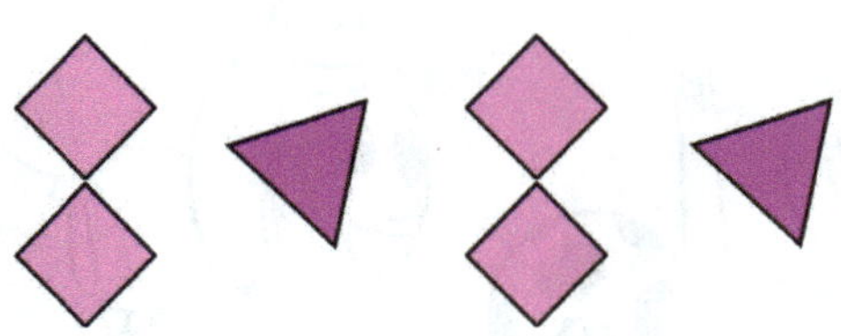

5. Add.

a.	**b.**	**c.**	**d.**	**e.**	**f.**
$2 + 2$	$1 + 4$	$9 + 0$	$1 + 3$	$2 + 3$	$8 + 2$

Solids

This is a **cube**. It is a box. Notice the squares in it.

This is a **sphere**. It is shaped like a ball.

A **cylinder** has a circle on the bottom and on the top.

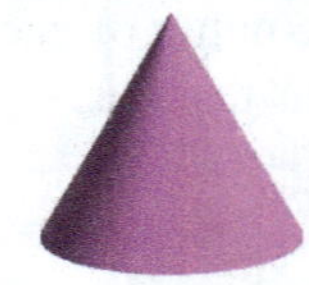

A **cone** has a pointed top and a rounded shape on the bottom.

A **vertex** is a point where two or more edges meet. It is a corner of a shape. The plural form of vertex is **vertices**.

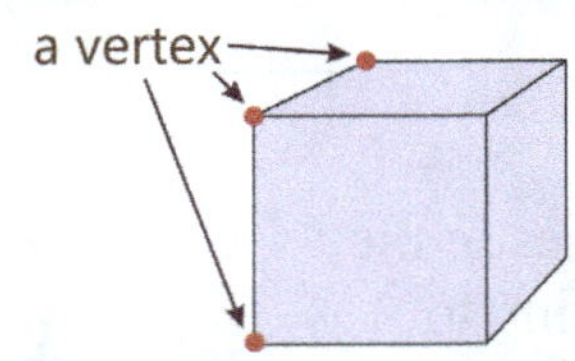

1. **Activity.** You need: a set of geometric solids (cubes, spheres, cylinders, and cones).

   First, check that the child knows the name of each solid. Then, hand the child two solids to compare. They could be two different solids, or the same type of solid but of different size and/or color.

   Ask questions such as "How are these shapes different from each other?", "How are these shapes alike?", and "Which shape is easier to roll?"

   For example, one can count how many vertices (corners) each solid has. One can describe the faces each solid has (e.g. the cube has squares, and the cylinder has circles). Also, one can describe the cone as being "pointy" and the sphere as being "round". The idea is to compare the solids using informal language.

2. Name each shape.

3. Count how many objects of each kind of shape are shown in the picture. Color the picture.

   cone _________

   sphere _________

   cube _________

   cylinder ________

2-D Shapes and 3-D Shapes

This square is a two-dimensional, or a "2-D" shape. Two-dimensional shapes are **flat**, like paper is flat.

They take up space in two directions (or dimensions): up and down and from side to side.

A cube is a three-dimensional, or a "3-D" shape. We call them **solids.**

Three-dimensional shapes take up space in three directions (or dimensions): up and down, from side to side, and from front to back.

1. **Activity:** You need: a set of geometric solids and a set of pattern blocks.

   Using shapes from both sets to illustrate, explain the basic differences between two-dimensional and three-dimensional shapes. You can use the teaching box above.

   Next, show the child one 2-D shape and one 3-D shape side-by-side (such as a circle and a sphere, or a cube and a square). Ask "How are they alike?", "How are they different?"

   Next, mix both sets of shapes together, and have the child sort the shapes (2-D in one pile and 3-D in another pile).

2. **Activity:** You need: a set of geometric solids.

   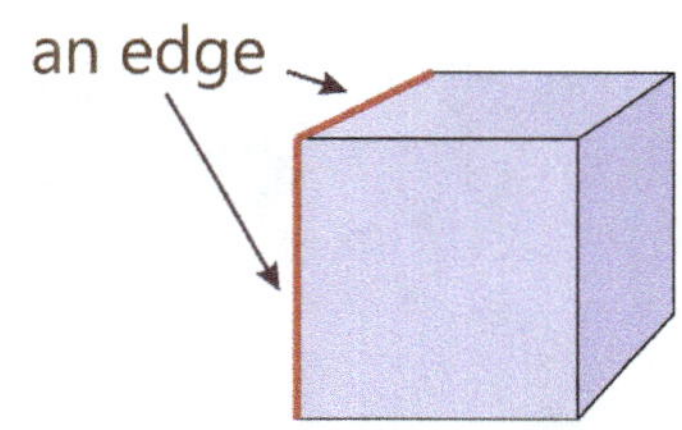

   Explain to the child what an **edge** is (a side in a 3-D shape that is like a line). Have the child trace edges in solids with their finger.

   Then, have the child count the number of edges in various solids.

3. **a.** Look at a cube (hold it). What 2-D shape can you find in it and where? Show your teacher.

   b. Look at a cylinder (hold it). What 2-D shape can you find in it and where? Show your teacher.

4. Are the shapes flat (2-D) or solids (3-D)? Draw a line from each shape to the correct answer.

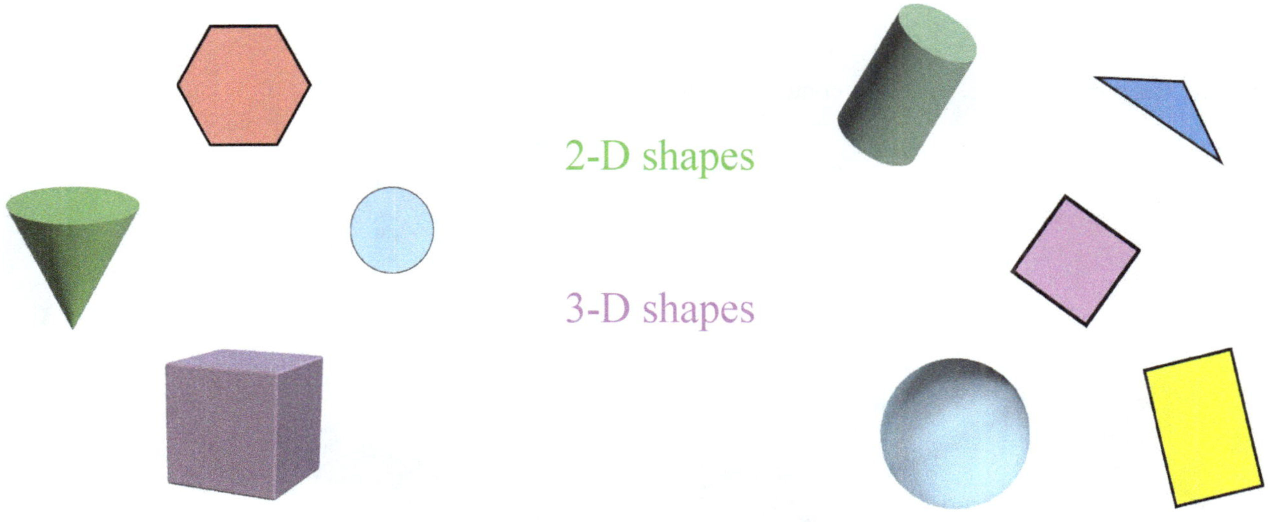

Shapes and Everyday Objects

1. **Activity:** Walk around your classroom, house, or some other location with your teacher, and look for objects that have the shape of a cube, sphere, cylinder, or cone.

   (Note to the teacher: You may need to specifically place items that have these shapes in the room for the student to find. For example, you could use boxes, oranges, cans of food, and party hats.)

   List here two things you found that are in the shape of a *cube*.

   _________________________ and ______________________________.

   List here two things you found that are in the shape of a *cylinder*.

   _________________________ and ______________________________.

2. Draw a line to match each object with the solid that is the same basic shape.

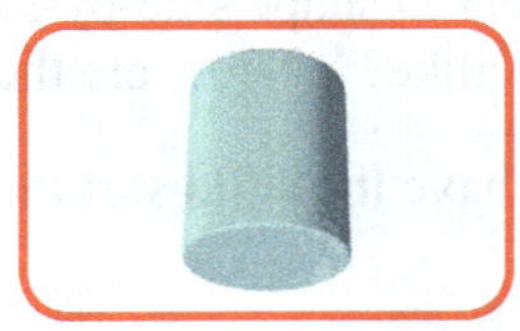

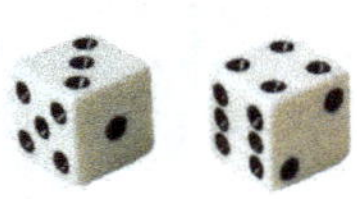
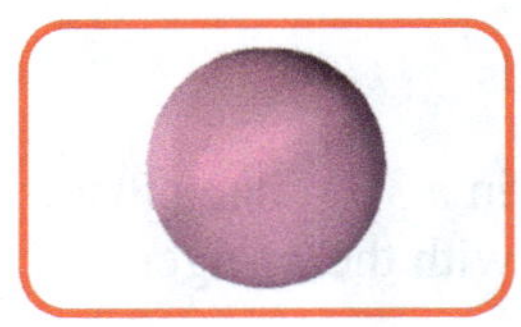

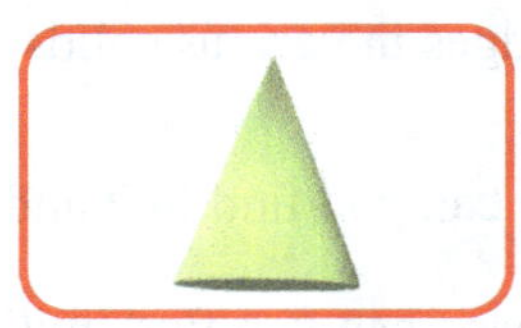

3. Name the basic shape (cube, cylinder, cone, or a sphere).

a. 
b. 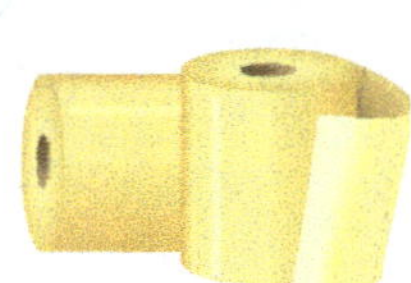
c. 
d. 

e. 
f. 
g. 
h. 

Building 3-D Shapes

1. **Activity.** You will need: modeling clay and toothpicks.

   Build different three-dimensional shapes using little clay balls and toothpicks. Build at least three different shapes. Below are some examples, but you can build more than these, including more complicated shapes and shapes that are a combination of several basic shapes (e.g. a cube and a pyramid on top of it, or two pyramids combined). Be imaginative and have fun!

   Tip 1: In many cases, you probably want to start out by building the bottom (or base) shape, such as a square or a triangle.

   Tip 2: Break some toothpicks in half to also have shorter sticks to work with.

   Questions:

   In these models, what corresponds to the **vertices** of a shape?

   What corresponds to the **edges**?

   (optional) Which of the shapes you built are flimsy and which are sturdy? Why?
   (Hint: It probably has to do with the presence or absence of triangles.)

This is a square pyramid. (It has a square as the bottom shape.)

This is a tetrahedron, which is a type of a triangular pyramid.

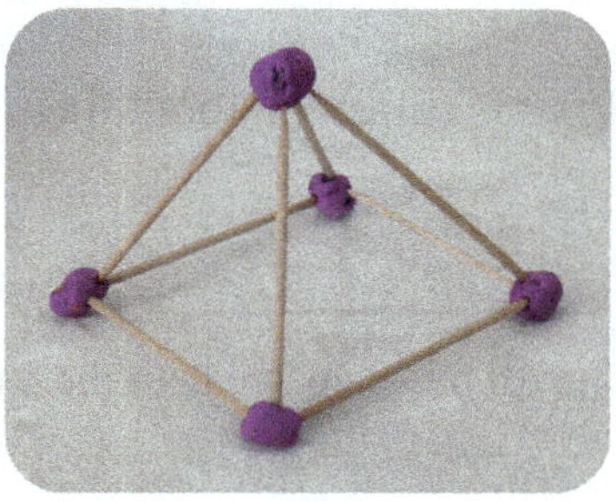

This is also a square pyramid.

You know what this shape is! It is not very stable when built out of sticks.

This is (trying to be) a box. I broke some toothpicks in half to make it.

This is a tent-shape. (In math, it is called a triangular prism.)

Comparing Shapes 1

What is same about these two shapes?
What is different?

The same:
They are both green.
They are both the same size (approximately).
They both have sides that are the same length.

Different:
One has four sides and the other has three.
Also, one has four corners and the other has three.
One is a square and the other is a triangle.

1. What is the same about these two shapes?

   What is different?

2. Are both of these shapes *triangles*?

   What is different about them?

3. Find all the shapes where the shape has sides of equal length.

a. **b.** **c.** **d.** **e.** **f.**

4. What is the same about these two shapes?

   What is different?

5. Mary said, "Both of these shapes are squares."
   Is she correct? Explain.

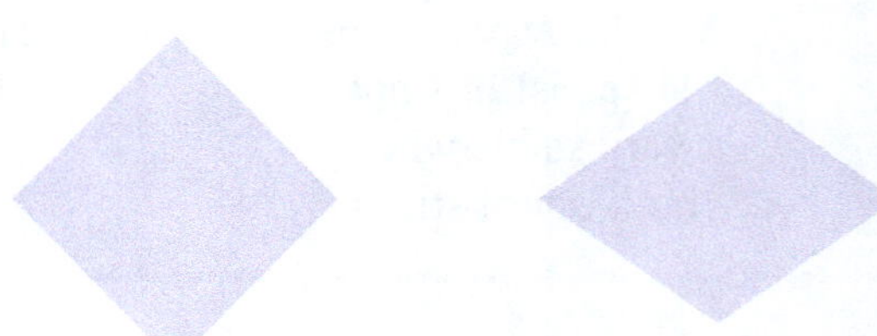

Comparing Shapes 2

A **vertex** is a point where two or more sides meet. It is a corner of a shape. The plural form of vertex is **vertices**.

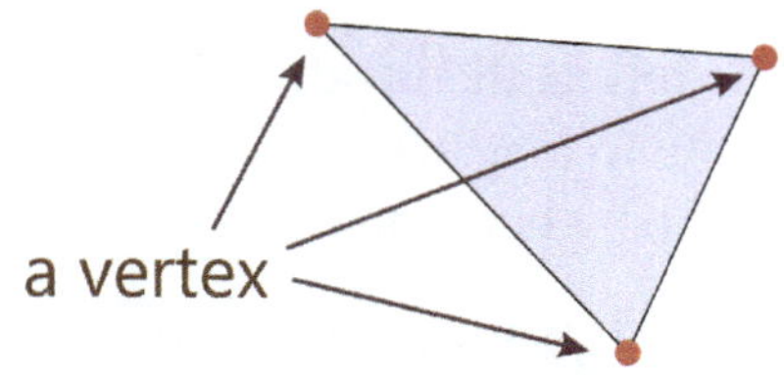

1. **Activity:** You need: a set of pattern blocks or a set of cut-out two-dimensional shapes.

   Explain to the child what a **vertex** is (a corner in a shape). Have the child count the number of vertices in various shapes.

   Then, ask the child to count the number of sides that various shapes have.

   Lastly, ask the child to check if a shape has the same number of sides and vertices. Use at least a triangle, a square, and a hexagon, but other polygons are great, too.

2. Tell or write how many vertices (corners) each shape has.

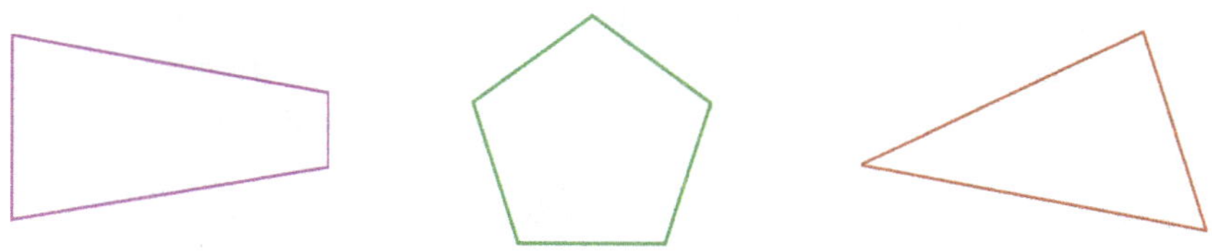

_______ _______ _______

3. Tell or write how many sides each shape has.

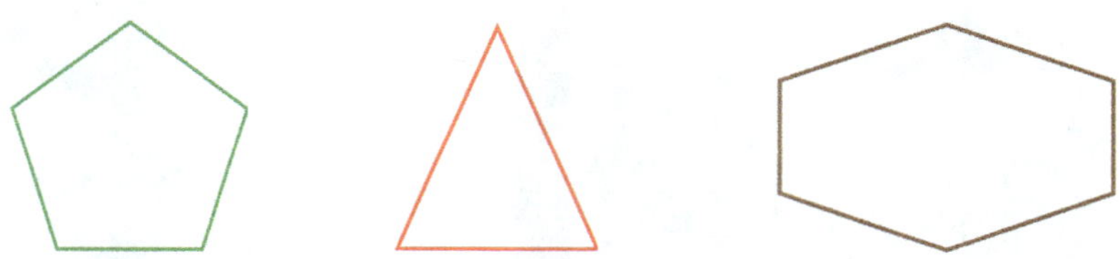

_______ _______ _______

4. Circle the solid with most vertices (corners).

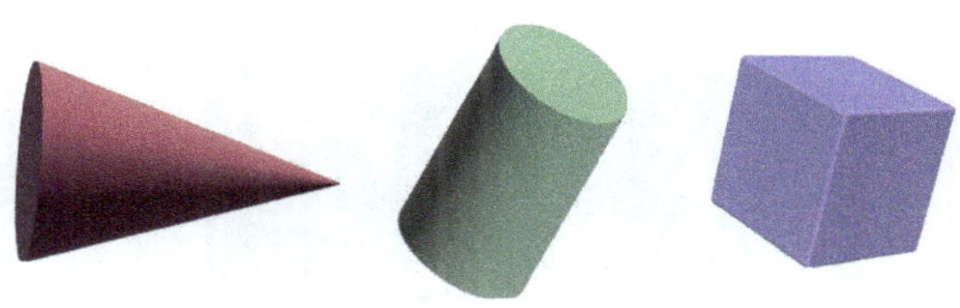

5. Alex said, “The difference between these two is, one is big and the other is small. And what is the same is, they both have four sides and they’re both orange.”

   Is he correct? Explain.

Position Words 1

Position words tell us the position of an object in relation to another object. In other words, they tell us where the object is.

Where is the cat? It is **on** the box.

Where is the box? It is **under** the cat and **beside** the ball.

Where is the ball? It is **beside** the box.

On, **under**, and **beside** are position words. Some other position words are **above**, **below**, **next to**, **in front of**, and **behind**.

1. **Activity.** Using objects in the basic shapes we have studied, plus others you have available, have the child model scenarios with position words. For example, if you give the child a cylinder and say "The cylinder is under the chair," the child then has to illustrate the sentence by placing the cylinder under a chair.

2. Circle the word that tells the position of each animal or object.

a.

The bunny is **on** / **beside** the couch.

b.

The cat is **in front of** / **under** the ball.

c.

The lamp is **behind** / **on** the table.

d.

The ball is **in** / **next to** the fridge.

Position Words 2

1. **Activity.** You will need: objects in the shapes we have studied (cube, cone, cylinder, sphere) positioned in the room in such a manner that you can describe their relative position to other objects. For example, place a cube-shaped box behind a chair, a party hat (a cone) in front of a closet, or a can of food beside a mug.

   Walk around with the child. Ask questions about the basic shapes of these objects, *and* ask the child to describe the position of those objects in relation to others. For example: "What shape is this box?", "Where is the box?" or "Where is the box in relation to the chair?" etc.

2. Tell your teacher the answers to the following questions.

a.

In what basic shape is the candle?

Where is it (in relation to the desk)?

In what basic shape is the ball?

Where is it (in relation to the desk)?

b.

In what basic shape is the red thing?

Where is it (in relation to the chair)?

In what basic shape is the trash can?

Where is it (in relation to the chair)?

3. Circle the correct position word.

a.

The cat is **behind / under** the table.

b.

The cat is **beside / behind** the ball.

4. Add and subtract.

a. 2 + 2 = _____	**b.** 1 + 4 = _____	**c.** 0 + 3 = _____
d. 5 − 2 = _____	**e.** 4 − 3 = _____	**f.** 3 − 2 = _____

Mixed Practice 5

1. Subtract. (The child may use objects or draw pictures to help.)

a.	**b.**	**c.**	**d.**	**e.**	**f.**
10	7	5	8	3	6
− 4	− 5	− 0	− 8	− 2	− 3

2. Tell whether each shape is a two-dimensional or three-dimensional shape.

a. **b.** **c.** **d.** **e.** **f.**

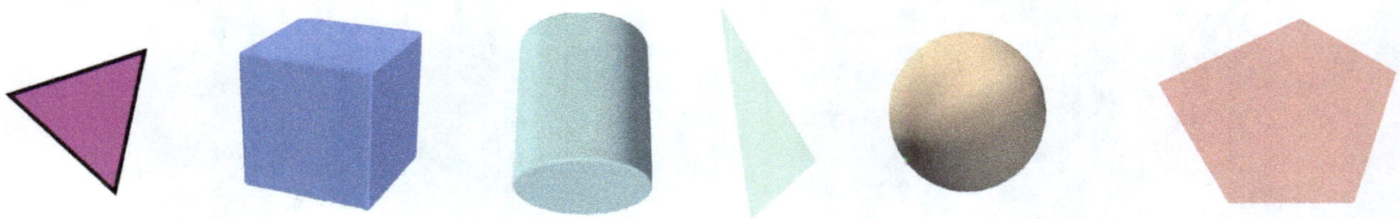

3. Number the objects in order from <u>lightest to heaviest</u> (1, 2, 3).

4. Would you do these activities at evening/night, in the morning, or in the afternoon? Tell your teacher.

feed your pet

go to bed

brush your hair

5. (optional) Cross out nine cents. How many are left? Write the answer.

Cross out nine cents.
How many are left?

____

Ordinal Numbers 1

Ordinal numbers are used when talking about the **order** of things.

first second third fourth fifth

1. The rabbits in this line are hungry!

a. Color **three** rabbits.

b. Color the **third** rabbit.

2. The curious baby elephants are following the men.

a. Color **two** elephants.

b. Color the **second** elephant.

3. **a.** Circle the fifth fish **from the right**.

b. Circle the first rabbit **from the left**.

c. Circle the fourth cat **from the right**.

Position Words 3

1. Tell your teacher the answers to these questions.

   a. In what basic shape is the red bottle?

   b. Where is it (in relation to the chair)?

   c. In what basic shape is the orange thing?

   d. Where is it (in relation to the chair)?

2. Draw pictures to illustrate the situations.

a. The butterfly is above the flower.	**b.** The shoe is beside the box.

3. Match each object to all words that accurately describe it.
(**Note:** Describing an object as "tall" or "short", "heavy" or "light", etc. can be very subjective, but that is fine in this exercise. The purpose here is for the child to use more than one word to describe the attributes of an object, not to absolutely correctly classify something as "tall", "short", "light", etc.)

heavy

small

long

tall

light

short

big

Ordinal Numbers 2

1. These chickens are going to the barn to visit the horses.

a. Color **four** chickens.

b. Color the **fourth** chicken (from the left).

2. Color the 2nd balloon **from the right** blue.

Color the 5th balloon **from the left** green.

Color the 3rd balloon **from the left** red.

Color the 1st balloon **from the left** purple.

Color the 4th balloon **from the right** yellow.

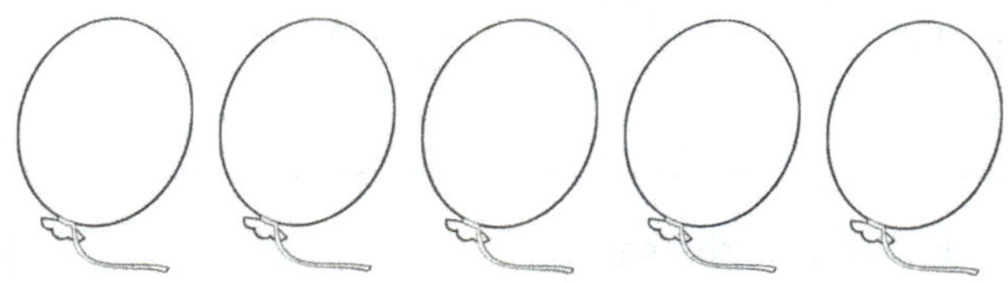

3. **a.** Draw an **X** in the first box **from the right.**

b. Draw an **X** in the third box **from the left.**

4. **a.** Circle the second flower **from the left.**

b. Circle the third starfish **from the right.**

c. Circle the fifth butterfly **from the right.**

Puzzle Corner

Using real dimes, nickels, and pennies, make these amounts of money:

28¢ 45¢ 33¢ 80¢

Comparing Shapes 3

1. Tell what is the same and what is different about each set of two shapes.
(The teacher can write down the answers, if desired, or you can do it orally only.)

a. What is the same: What is different:		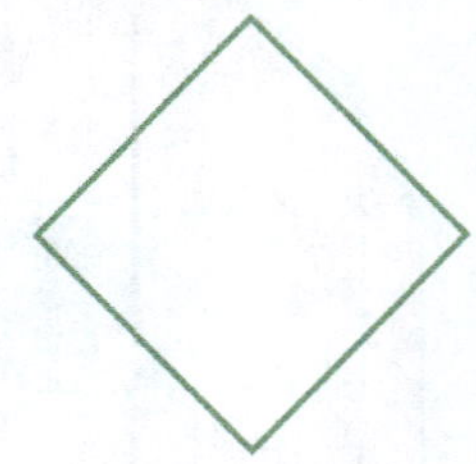
b. What is the same: What is different:	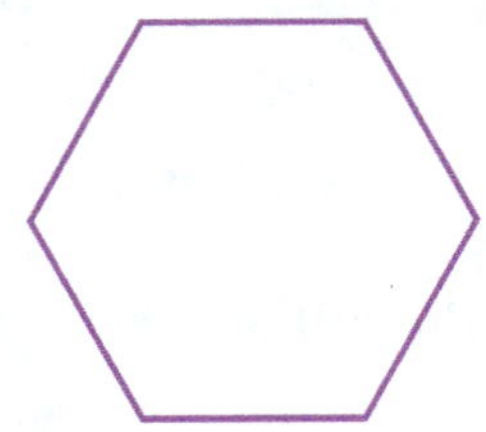	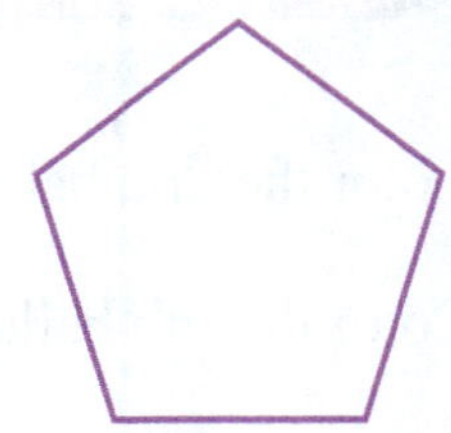
c. What is the same: What is different:		
d. What is the same: What is different:		

2. Draw a line from each shape to the correct answer.

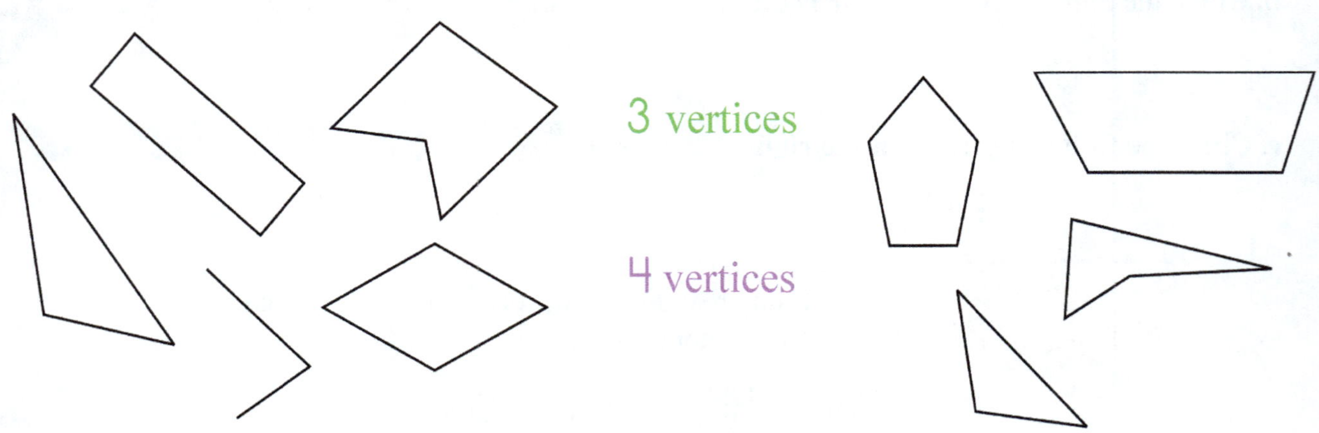

Drawing Shapes 2

1. In the boxes below, sketch (approximate drawing is sufficient) the indicated shapes.

a. a big green triangle	**b.** a small blue triangle	**c.** a small rectangle
d. a red circle	**e.** a brown square	**f.** a square that is tilted

2. Sketch the basic shape(s) that you see in each picture (square, rectangle, circle, hexagon).

a.

b.

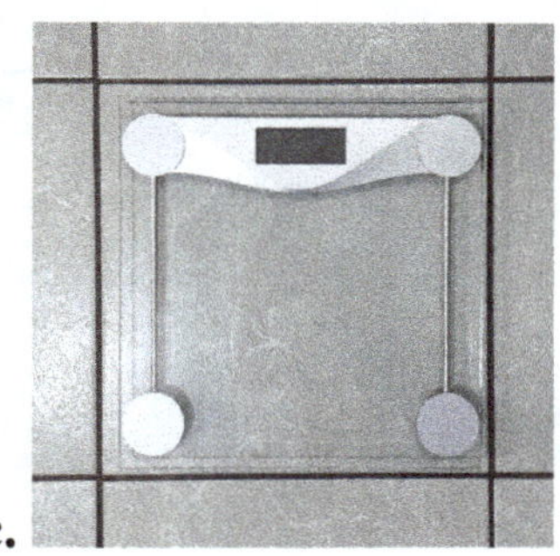

c.

Shapes and Order

1. Answer (tell your teacher).

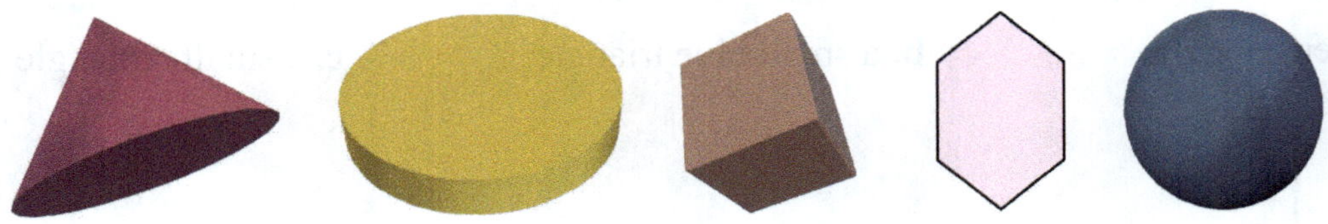

a. Find the the fifth shape from the right. What kind of shape is it?

b. Find the second shape from the left. What kind of shape is it?

c. What kind of shape is the third shape from the right?

d. What kind of shape is the fourth shape from the left?

e. What kind of shape is the first shape from the right?

2. Tell or write how many corners each shape has.

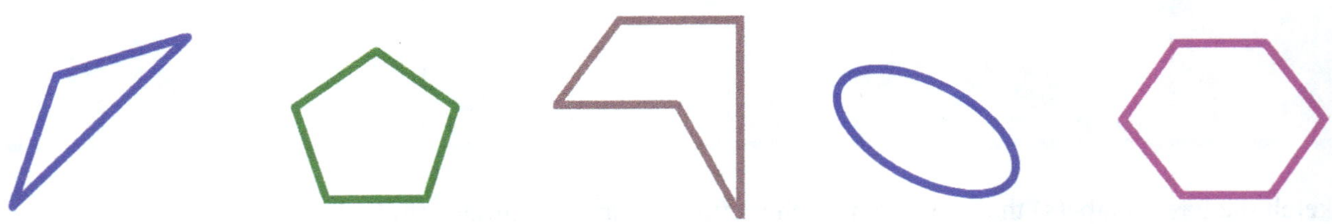

a. ________ **b.** ________ **c.** ________ **d.** ________ **e.** ________

3.

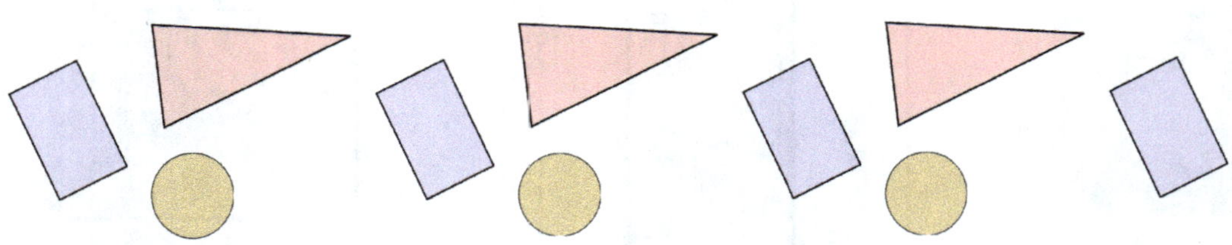

a. Circle the fourth rectangle from the right.

b. Circle the second circle from the left.

c. Circle the first triangle from the right.

4. Add.

a. $\begin{array}{r} 5 \\ +\ 0 \\ \hline \end{array}$ **b.** $\begin{array}{r} 2 \\ +\ 3 \\ \hline \end{array}$ **c.** $\begin{array}{r} 3 \\ +\ 1 \\ \hline \end{array}$ **d.** $\begin{array}{r} 2 \\ +\ 2 \\ \hline \end{array}$ **e.** $\begin{array}{r} 4 \\ +\ 1 \\ \hline \end{array}$ **f.** $\begin{array}{r} 1 \\ +\ 3 \\ \hline \end{array}$

Composing Shapes 1

Activity. You will need: a set of pattern blocks. You can use a commercial set made of plastic or foam. Or, you can print and cut out the blocks on the following page, or download the page from here:

https://www.mathmammoth.com/download/pattern-blocks.pdf

1. Find a hexagon in the set. But now, make a hexagon using two or more of the *other,* smaller shapes. There are several ways to do this! How many ways can you find?

2. The set has a triangle, but your task is to make a triangle using several shapes put together. It can be a big triangle! Again, find several ways to do it.

3. Make a rectangle. Can you make several different ones?

4. Use several shapes to make these shapes:

a. 

b. 

c. 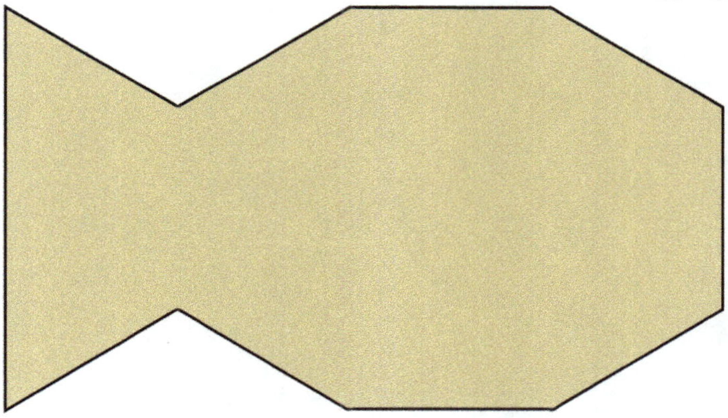

5. (challenge) Use several shapes to make this shape. **Note:** The image on the right is *not* to scale. It is half the size of the real shape.

(This page is intentionally left blank.)

(This page is intentionally left blank.)

Composing Shapes 2

1. **Activity.** Cut out the shapes at the bottom of this page. Alternatively, trace them on another paper and then cut them out. Or, you can download and print this page from here:

https://www.mathmammoth.com/download/composing-shapes.pdf

a. Make a rectangle by putting together several of the shapes.

b. Now make a bigger rectangle.

c. Make a square.

d. Make a triangle by putting together several shapes.

e. Make another, different triangle. How big a triangle can you make?

f. Make a diamond-shape.

g. Make any shape you want! Have fun!

2. If you put together these shapes, what new shape will you get?

a.
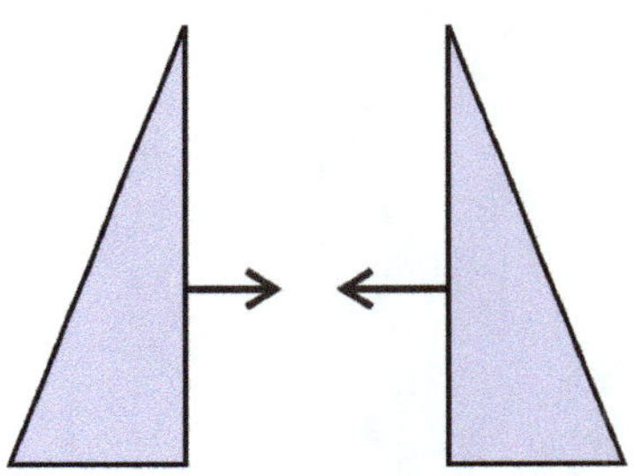

b.
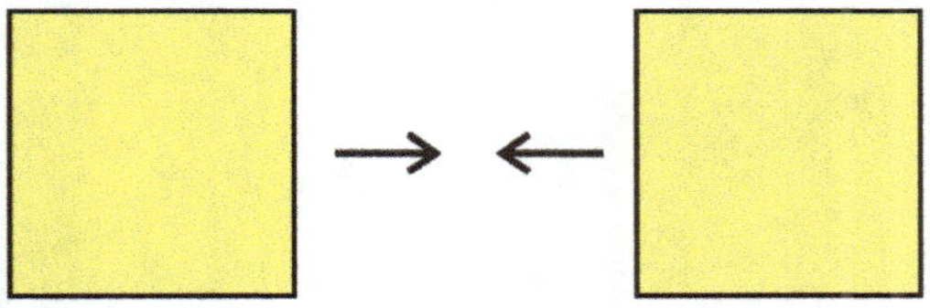

c.
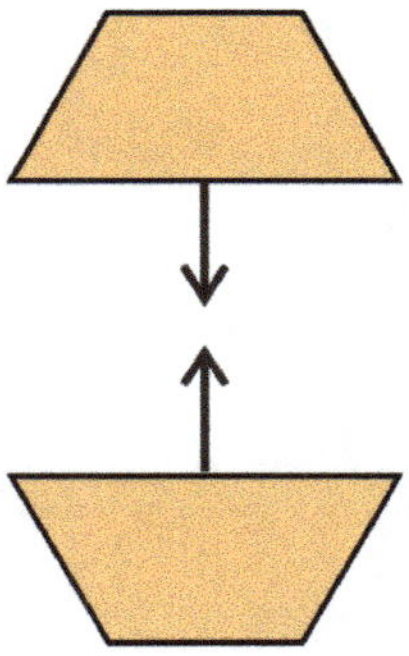

Cut out these:

(This page is intentionally left blank.)

Mixed Review Chapter 7

1. Count aloud **a.** from seventy-six to one hundred; **b.** from ten to thirty.

2. Count by tens from ten to one hundred.

3. Say the number word for these amounts.

a. 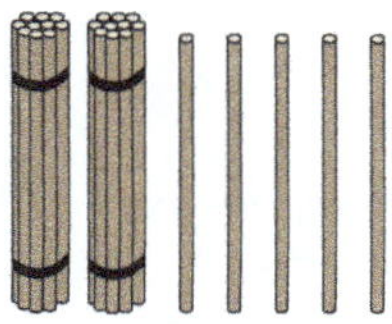

b. 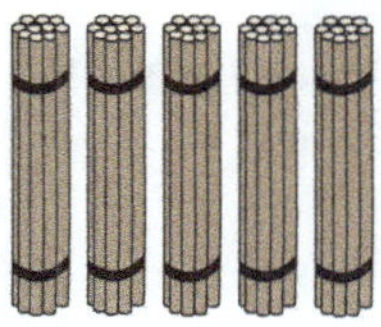

c. 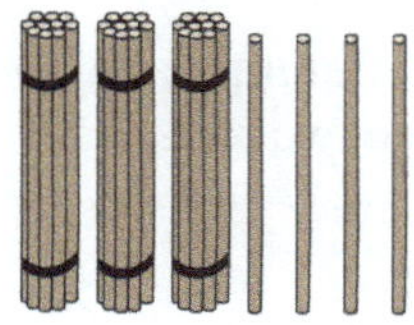

d. 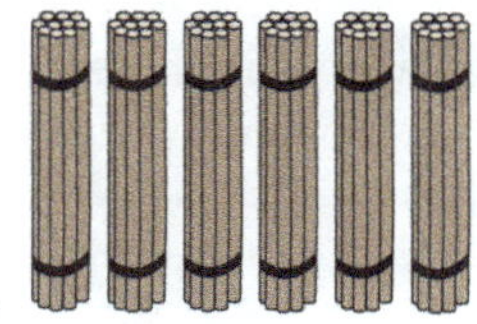

4. What is the same and what is different about these two shapes?
(The teacher can write down the answers, if desired, or you can do it orally only.)

5. Count the objects and write the number on the line.

a.  ______

b.  ______

6. Add and subtract. Notice: each box has a pattern that can help you!

a.	**b.**	**c.**
$10 - 1 =$ ______	$7 - 2 =$ ______	$3 + 2 =$ ______
$9 - 1 =$ ______	$6 - 2 =$ ______	$4 + 2 =$ ______
$8 - 1 =$ ______	$5 - 2 =$ ______	$5 + 2 =$ ______
$7 - 1 =$ ______	$4 - 2 =$ ______	$6 + 2 =$ ______

7. Subtract.

a. $4 - 2 =$ ____	b. $5 - 2 =$ ____	c. $3 - 1 =$ ____
$5 - 3 =$ ____	$4 - 4 =$ ____	$3 - 2 =$ ____

8. Draw dots in the empty circles, in different ways, so that the total is always 9.
(Find different ways to decompose 9 into two parts.)

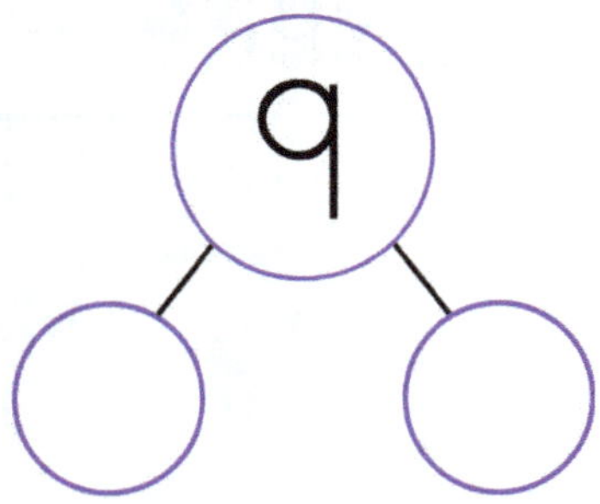

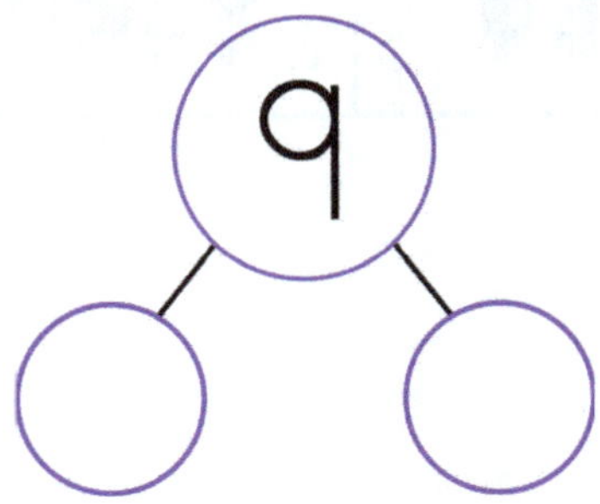

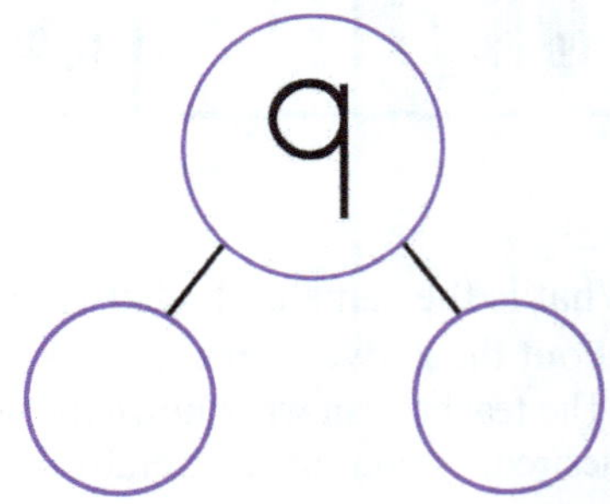

9. (optional) How many cents? Write the total on the line.

a.  ________¢

b.  ________¢

10. Choose a color for each kind of shape, and color them!

Triangles are ______________. Circles are ______________.

Squares are ______________. Rectangles are ______________.

Other four-sided shapes are ______________.

Chapter 7 Review

1. How many **vertices (corners)** does each shape have? Write it on the line.

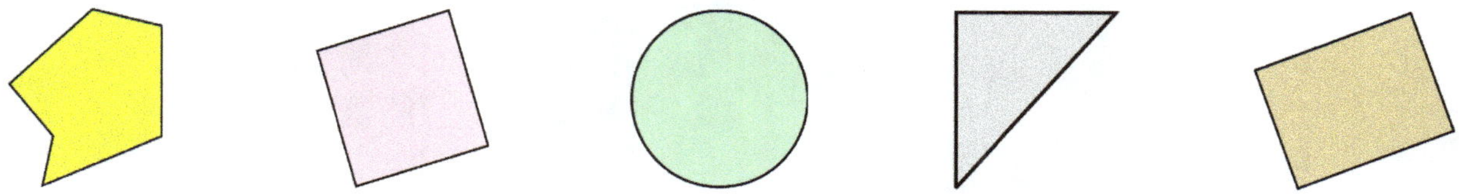

________ ________ ________ ________ ________

2. What is the same and what is different about these two shapes? Tell your teacher.

3. If you put together these shapes, what new shape will you get?

a.

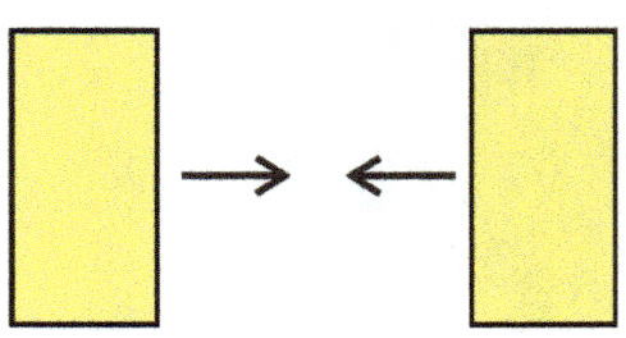

b.

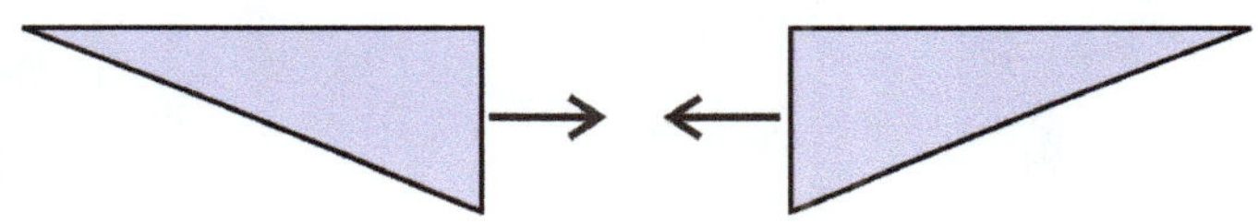

4. **a.** Circle the second cow from the right. **b.** Cross out the fifth cow from the left.

5. Tell your teacher the answers to the following questions.

a. In what basic shape is the box?

b. Where is it (in relation to the fridge)?

c. In what basic shape is the trashcan?

d. Where is it (in relation to the fridge)?

Chapter 8: More Addition and Subtraction
Introduction

In this chapter, we circle back to addition and subtraction, and also review two-digit numbers and counting by tens. There are no new concepts in this chapter, which means you can include lessons from here when studying earlier chapters, or omit the chapter (or some of its lessons) if time does not allow.

Don't forget to check out the worksheet maker for extra worksheets for these topics:
https://www.mathmammoth.com/private/Make_extra_worksheets_K.htm

Pacing Suggestion for Chapter 8

Feel free to take a day at times to just do activities and play games, without any written work.

The Lessons in Chapter 8	page	span	suggested pacing	your pacing
Addition & Subtraction Review	101	*1 page*	1 day	
Addition Equations 1	102	*1 page*	1 day	
Addition Equations 2	103	*1 page*	1 day	
Addition Equations 3	104	*1 page*	1 day	
Subtraction Equations 1	105	*1 page*	1 day	
Subtraction Equations 2	106	*1 page*	1 day	
Mixed Practice 6	107	*1 page*	1 day	
More Practice with Numbers	108	*2 pages*	1 day	
More Equations	110	*1 page*	1 day	
Word Problems: More Practice	111	*1 page*	1 day	
Mixed Review Chapter 8	112	*2 pages*	2 days	
Chapter 8 Review	114	*1 page*	1 day	
TOTALS		*14 pages*	13 days	

Games

These games are copied here from the earlier chapter introductions.

Some Went Hiding *(Decomposing numbers)*

You need: Up to 10 small objects, such as marbles, pennies, beans, or rocks.

Game play: Choose a target number, such as 7. The first player shows that many objects to the next player, and then quickly hides *some* of them behind his back (without showing how many). He shows the rest of them to the next player, who has to say how many "went hiding." If the player gives the right answer, it is then his turn to hide some and ask the next player to answer. If he gives a wrong answer, he forfeits his turn. This game appeals best to young children.

Variation: Instead of getting a turn to hide objects, the player who answers correctly may gain points or other rewards for the right answer.

Mammoths & Ladders with 10-frames *(Decomposing 10)*

You need: Mammoths & Ladders gameboard, or some other board game where you advance a marker. A marker for each player. A deck of 10-frame cards. Download printable 10-frame cards here:

https://www.mathmammoth.com/download/ten-frames-cards.pdf

Game play: Place the deck of 10-frame cards face down on the table. On your turn, draw one of those cards, and say the number that makes 10 together with the quantity shown on the card. For example, if the card shows 7, you're supposed to say "3". Then move your marker the number of steps you just said (three steps, in this case). Otherwise, follow the rules of the Mammoths & Ladders game (or the board game you chose to use).

Addition Challenge *(Addition, comparing numbers)*

You need: Number cards with numbers from 1 to 5. (Regular playing cards will work.)

Game play: In each round, each player is dealt two cards face up, and calculates the sum. The player with the highest sum gets all the cards from the other players. After enough rounds have been played to use all of the cards, the player with the most cards wins. If two or more players have the same sum, those players get an additional two cards and use those to resolve the tie.

Note: when using regular playing cards, children can count the symbols on them to find the answer. (Encourage them to count *on* from the bigger number, and not start at 1.)

Variation: Instead of cards with numbers 1-5, use cards with 1-6 or some other range.

Animal Rescue with Addition *(Addition)*

You need: Two six-sided dice. Or, if you have them, two four-sided dice will make for an easier game. Number cards could also be used. About two dozen stuffed animals, plastic animals, lego figurines, bear counters, or similar things. In a pinch, use beans or marbles, and ask the child(ren) to imagine they are some type of animals. An environment for the animals where they are "in danger." For example, this could be a blanket that represents seawater, or it could be something representing a barn on fire, etc.

Game play: On your turn, roll the two dice, and "rescue" as many animals as is the total of the dice. Those animals become yours. In the end of the game, the player with the most rescued animals wins.

Domino Crabs *(Addition)*

You need: Seven markers for each of the two players. A set of dominoes from which you remove the ones where the sum of the dots is 7 or more. (Keep the dominoes where the sum is from 0 to 6.)
Download the game board from here:

https://www.mathmammoth.com/download/Domino-Crabs.pdf

Game play: Place the dominoes face down as if in a pond in between the players. Place the game board in between the players also. On your turn, draw one domino from the pond and add the dots. Then place a marker at your number on the board that is the sum. If you cannot do so, you will skip your turn. The first player to place all their markers on the crabs is the winner.

Addition Go Fish *(Addition)*

You need: The deck of cards for this game, which you can download from here:
https://www.mathmammoth.com/download/Addition-Go-Fish.pdf

Game play: Deal each player four cards. Spread the rest of the cards face down on the table as if a "pond." On your turn, first try to find matches (such as 5 + ● and 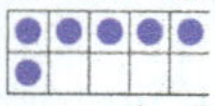) within your own cards. Remove any matches, placing them in your personal pile. Then, you may ask the player on your left for one specific number, such as, "Do you have an eight?" If the player has it (or one of the representations of 8), they have to give it. Otherwise, they say, "Go fish," and you will then take one card from the pond.

If you now get a match, you will again place those cards in your personal pile. Your turn will continue, and you can again ask the player on your left for one specific number. Once you cannot make a match with the cards, it is the next player's turn.

Continue playing until the pond is used up and no one can make any more matches. The winner is the player with the most cards in their personal pile.

Zebra Dash Addition *(Addition within 0-5)*

You need: A six-sided die and a marker for each player. Download the game board from here:

https://www.mathmammoth.com/download/Zebra-dash-addition.pdf

Game play: First, each player rolls the die, and the player with the highest number starts the game. On your turn, roll the die, and move your marker that many spaces. If that square has instructions on what to do, follow those. If it has an addition problem, answer that. If your answer is wrong, move your marker backward one space.

Every time you come to a square with a math problem, follow the same process: say aloud the answer, and if your answer is wrong, move your marker backward one space. And if it has other instructions, follow those. For example, let's say you answer 1 + 3 wrong, and go backward one space. That square states you move ahead one step, so you do so, and then you answer 1 + 3 again.

The person who reaches the watering hole first is the winner.

10 Out (or *5 Out* or *6 Out, etc.*) *(Number bonds)*

Note: This game is optional. It is challenging for kindergarten level. You can make it easier by providing manipulatives/objects or instructing the players to use their fingers.

You need: Lots of number cards from 1 to the target number.

Preparation: Choose a target number (sum) for the game, such as 10. Deal five cards to each player. Place the rest face down in a pile in the middle of the table.

Game play: On your turn, you may discard the card 10 (the target number) or any two cards in your hand that add up to it. You may also ask for one card from the player to your right (as in "Go Fish"), and if that player has it, they must give it to you. If you can discard all cards from your hand this way, you're the winner and the game ends. If not, take one card from the deck, and that ends your turn.

Variations:
* Deal more than five cards.
* Deal fewer cards if there are a lot of players or the players are very young.
* Allow players to discard *three* cards that add up to the target number.

Subtraction Icebergs *(Subtraction)*

This is a two-player game.

You need: Eight counters for each of the two players. A ten-sided die or one six-sided die and number cards from 1-10. Download the game board here:

https://www.mathmammoth.com/download/Subtraction-icebergs.pdf

Game Play: Each player places their eight counters on their eight icebergs. On your turn, to get two numbers to subtract, roll the ten-sided die twice, or roll the six-sided dice and draw one number card. Subtract the two numbers. Remove the counter from the iceberg that shows the answer to the subtraction. If you cannot go, you skip your turn. The player that removes all their counters from the icebergs first is the winner.

Subtraction Challenge *(Subtraction)*

You need: A deck of number cards from 1-10. You might also remove 10s, 9s, and 8s, using your judgment, in which case you may need two decks. Alternatively, a set of dominoes works well also. Limit it to dominoes with a maximum of 10 dots on one side.

Game Play: In each round, each player is dealt two cards (or 1 domino) face up, and has to calculate the difference (subtract). The player with the highest difference gets all the cards (dominoes) from the other players. After enough rounds have been played to use up the deck, the player with the most cards wins.

If two or more players have the same difference, then those players get an additional two cards (one domino) and use those to resolve the tie.

Let's Make a Difference! *(Subtraction)*

You need: A large deck of number cards with numbers 1-10. Optionally, some small objects, number path, or an abacus to help children to do the subtractions.

Preparation: Deal six cards to each player. If a player cannot hold the cards, they can place them in front of them on the table. Place the rest face down in a pile in the middle of the table.

Game play: On your turn, check if you can make a subtraction sentence using any three cards in your hand. If so, say that subtraction aloud, and put those three cards in your personal pile. If not, take *one* card from the middle pile, and check again. If you still cannot make a subtraction, your turn ends. At the end of your turn, draw enough cards from the middle pile so that you once again have at least six cards in your hand (if you already have six or more, don't draw more at this point). The turn then passes to the next player.

Play in this manner until there are no more cards in the middle pile. At that point, the player with the most cards in their personal pile wins.

Note: Essentially, players can check if they have three cards that make an *addition* sentence. However, it is best if the child notices that on their own.

Variation: For an easier game, remove some of the higher numbers from the deck. You can also deal each player five cards, instead of six; however this will make it less likely that the player can make a subtraction sentence.

Zebra Dash Subtraction *(Subtraction within 0-5)*

You need: A six-sided die and a marker for each player. Download the game board from here:

https://www.mathmammoth.com/download/Zebra-dash-subtraction.pdf

Game play: First, each player rolls the die, and the player with the highest number starts the game. On your turn, roll the die, and move your marker that many spaces. If that square has instructions on what to do, follow those. If it has a subtraction problem, answer that. If your answer is wrong, move your marker backward one space.

Every time you come to a square with a math problem, follow the same process: say the answer aloud and if your answer is wrong, move your marker backward one space. And if it has other instructions, follow those. For example, let's say you answer $5 - 2$ wrong, and go backward one space. That square states you move ahead one step, so you do so, and then you answer $5 - 2$ again.

The person who reaches the watering hole first is the winner.

Keep Your Card *(Numbers from 11 to 20)*

You need: A deck of number cards from 1 through 9. One number card with 10.

Game Play: Shuffle the deck and place it face down between the players. Place the card with 10 face up next to the deck. On your turn, draw one card from the deck. Add that number and 10, and say the sum out loud. Show your card to everyone. If you added correctly, you get to keep the card, so place it in your personal pile. If not, the card goes back to the bottom of the deck. Play until the deck is used up. The winner is the player with most cards at the end of the game.

Variation: (challenge) Instead of having each player draw just one card from the deck, deal three cards to each player in the very beginning. On your turn, check if you can add two of your cards and get a sum that is more than 10. If yes, do so, and if your sum is correct, place the two cards into your personal pile. End your turn by drawing two cards from the deck, so that you have three cards again in your hand.

If your sum is not correct, those two cards go back to the bottom of the deck, and you draw two more from the deck. If you cannot find two cards that add up to something more than ten, you will draw one more card from the deck, and pass your turn.

See also the games and activities in the chapter 6 introduction.

Math Mammoth Practice Zone

Addition Hidden Picture Game
Solve addition problems within 0-6 while uncovering a hidden picture in this fun game that children really like!
https://www.mathmammoth.com/practice/mystery-picture#min=0&max=6

Subtraction Hidden Picture Game
Solve subtraction problems within 0-5 while uncovering a hidden picture. A fun game that children really like!
https://www.mathmammoth.com/practice/mystery-picture-subtraction#min=0&max=5

"7 Up" card game
You will see seven cards dealt face up. Choose any two cards that make 10 (or your chosen sum) to discard.
https://www.mathmammoth.com/practice/seven-up

Number Bonds
This colorful online activity or game allows a child to practice the concept of number bonds. The child can either work with pictures (jewels) or with numbers.
https://www.mathmammoth.com/practice/number-bonds

Further Resources on the Internet

These resources match the topics in this chapter, and offer online practice, online games (occasionally, printable games), and interactive illustrations of math concepts. We heartily recommend you take a look. Many people love using these resources to supplement the bookwork, to illustrate a concept better, and for some fun. Enjoy!

https://links.mathmammoth.com/Kch8

Scan me

Addition & Subtraction Review

1. Subtract.

a. $\begin{array}{r}5\\-4\\\hline\end{array}$ **b.** $\begin{array}{r}2\\-0\\\hline\end{array}$ **c.** $\begin{array}{r}3\\-3\\\hline\end{array}$ **d.** $\begin{array}{r}4\\-2\\\hline\end{array}$ **e.** $\begin{array}{r}5\\-0\\\hline\end{array}$ **f.** $\begin{array}{r}3\\-1\\\hline\end{array}$

2. Solve the problems. You can draw pictures to help you.

a. Mara has five limes and Kim has . They put their limes together. How many limes do they have in total?

b. Aaron and Bill were playing with seven balloons in the yard. Three balloons blew away. How many balloons are left?

3. Decompose 7 into pairs in different ways. Draw two groups of fish that make 7, and write the addition.

a.

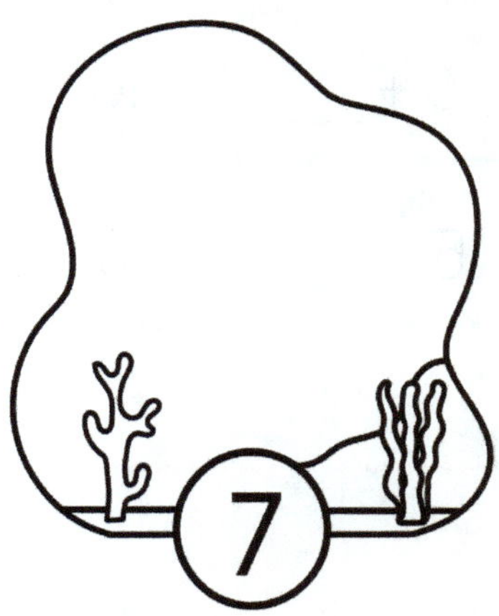

___ + ___

b.

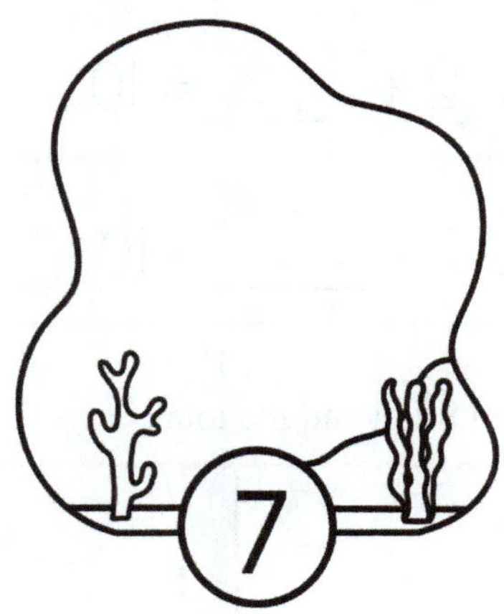

___ + ___

c.

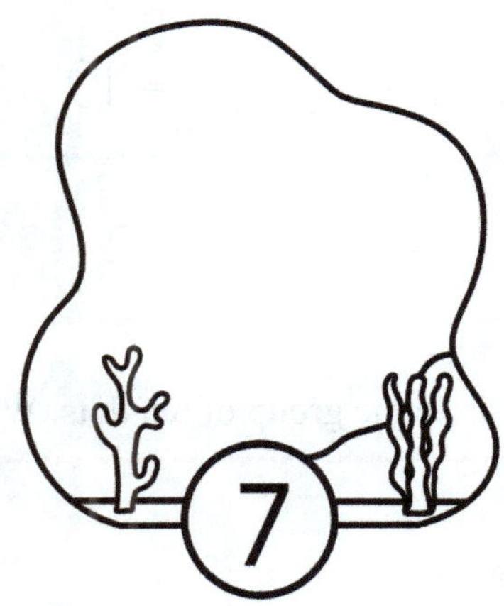

___ + ___

4. Add.

a. $\begin{array}{r}3\\+0\\\hline\end{array}$ **b.** $\begin{array}{r}1\\+1\\\hline\end{array}$ **c.** $\begin{array}{r}2\\+3\\\hline\end{array}$ **d.** $\begin{array}{r}4\\+1\\\hline\end{array}$ **e.** $\begin{array}{r}0\\+1\\\hline\end{array}$ **f.** $\begin{array}{r}7\\+1\\\hline\end{array}$

Addition Equations 1

1. Play one of the addition games from the chapter introduction.

2. Draw dots (or other shapes) to show these addition equations.

a. 3 + 7 = ____	b. 1 + 6 = ____	c. 4 + 4 = ____

3. Draw more animals so that the total is 10. Then write an addition.

a.

____ + ____ = 10

b.

____ + ____ = ____

4. Find the missing number so that the total is 10. The ten-frame can help you.

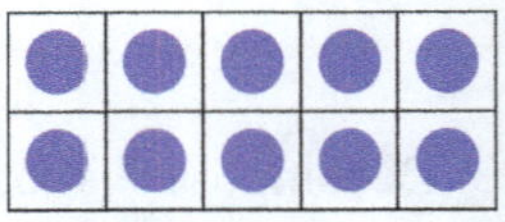

a. 4 + ____ = 10	b. 2 + ____ = 10	c. 6 + ____ = 10
d. 1 + ____ = 10	e. 7 + ____ = 10	f. 5 + ____ = 10

5. Circle the group of ten dots. Write an addition and the total.

a.

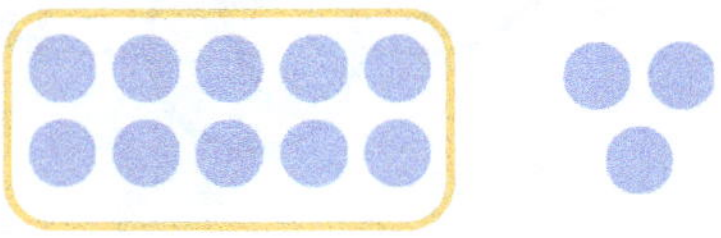

10 + __3__ = ____

b.

10 + ____ = ____

c.

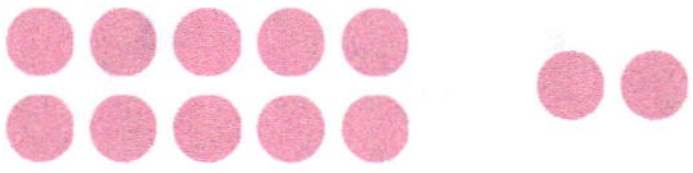

____ + ____ = ____

d.

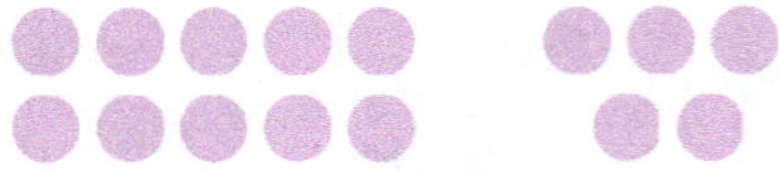

____ + ____ = ____

Addition Equations 2

1. **Game.** Play Mammoths & Ladders with ten-frames, variation 2 (see the chapter introduction).

2. Draw more dots of a different color to make 10. Write the missing numbers in the additions.

a. 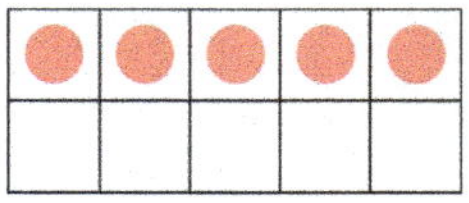

_____ + _____ = 10

b. 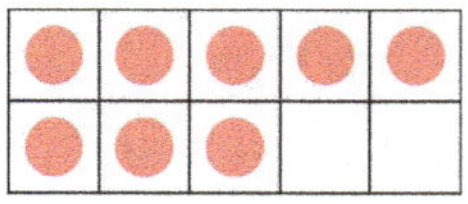

_____ + _____ = 10

c. 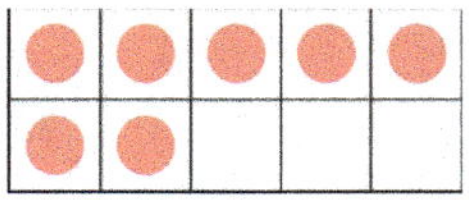

_____ + _____ = 10

d. 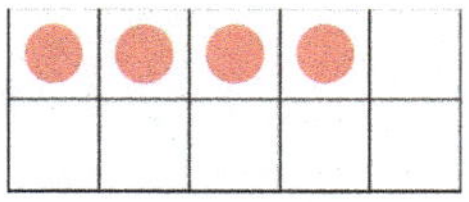

_____ + _____ = 10

3. Circle the group of ten dots. Then write a sum and the total.

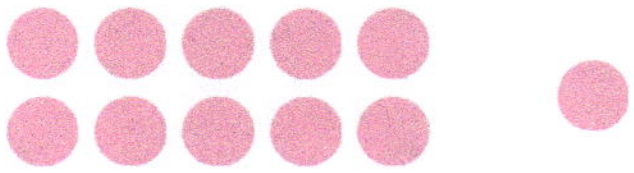

a. ______ + _____

____

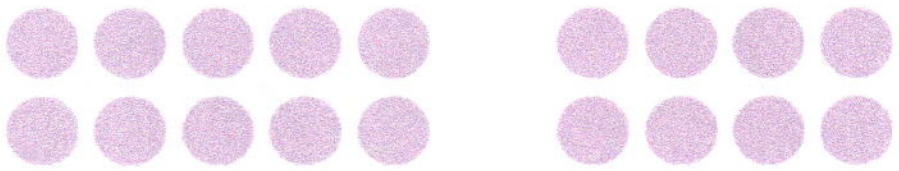

b. ______ + _____

____

4. Write each number as a sum of 10 and some other number. Draw dots to illustrate the equation.

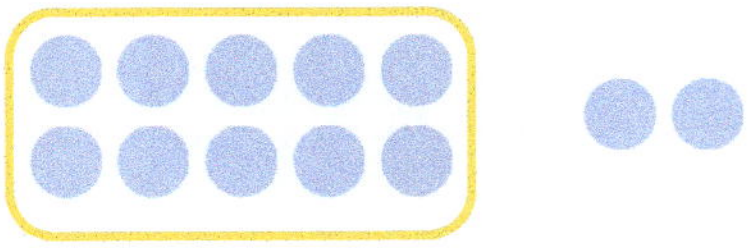

a. 12 = ______ + _____

b. 17 = ______ + _____

c. 14 = ______ + _____

d. 19 = ______ + _____

Addition Equations 3

1. Draw two groups of dots that total nine, in different ways. Write the missing numbers in the equations.

a. 9 = _____ + _____	**b.** 9 = _____ + _____
c. 9 = _____ + _____	**d.** 9 = _____ + _____

2. Write each number as a sum of 10 and some other number. Draw dots to illustrate the equation: draw a group of 10 dots, and some more dots.

a. 13 = _______ + _____	**b.** 16 = _______ + _____
c. 15 = _______ + _____	**d.** 18 = _______ + _____

3. Solve. You can draw or use objects to help you solve the problems.

a. Tanya wants to give ten flowers to her mom. She has already picked three flowers. How many more does she need to pick?

b. Erica picked ten flowers. She gave her mom six of them. How many more does she have left now?

Subtraction Equations 1

1. Cross out the indicated number of objects. Write a subtraction.

Cross out two.	Cross out five.
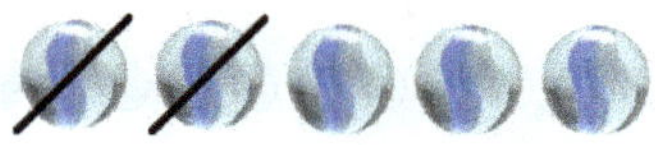	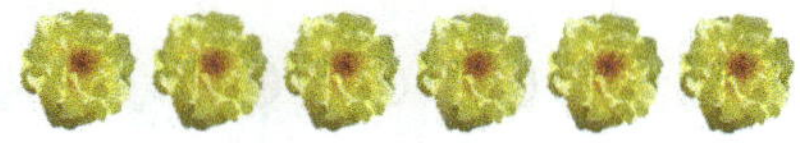
a. _____ − _____ = _____	**b.** _____ − _____ = _____

2. Draw something to illustrate the subtractions. You can also use objects.

a. $7 - 3 =$ _____	**b.** $10 - 5 =$ _____
c. $8 - 6 =$ _____	**d.** $9 - 7 =$ _____

3. Solve. You can draw pictures to help.

a. Emily had ten pennies. Then she lost  .
How many pennies does she have left?

b. Alex has two dogs. He also has cats named Kitty, Leo, and Bella.
How many pets does Alex have in total?

4. Subtract.

a. $\begin{array}{r} 4 \\ -0 \\ \hline \end{array}$ **b.** $\begin{array}{r} 2 \\ -2 \\ \hline \end{array}$ **c.** $\begin{array}{r} 4 \\ -3 \\ \hline \end{array}$ **d.** $\begin{array}{r} 5 \\ -4 \\ \hline \end{array}$ **e.** $\begin{array}{r} 3 \\ -1 \\ \hline \end{array}$ **f.** $\begin{array}{r} 5 \\ -5 \\ \hline \end{array}$

Subtraction Equations 2

1. **Game.** Play one of the subtraction games (see the chapter introduction).

2. Cover the *darker* dots and complete the subtraction.

a. 

7 – ____ = ____

b. 

____ – ____ = ____

3. Now we will write two subtractions. First, cover the darker flowers and write a subtraction. Then, cover the lighter flowers instead, and write another subtraction.

a. 8 – ____ = ____

8 – ____ = ____

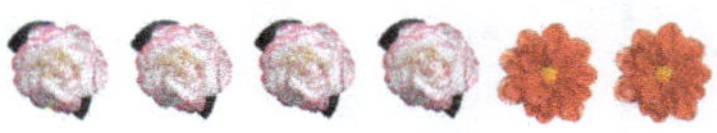

b. ____ – ____ = ____

____ – ____ = ____

4. Solve. You can draw pictures to help.

a. Four turtles were swimming in the pond.
Three more turtles joined them.
How many turtles are in the pond now?

b. Cheryl's mom cut a pizza in eight slices.
Oh no! Cheryl's dog Skippy ate two
slices! How many slices of pizza are left?

5. Subtract. Notice the patterns.

a. $5 - 0 =$ ____	**b.** $4 - 3 =$ ____	**c.** $3 - 1 =$ ____
$5 - 1 =$ ____	$4 - 2 =$ ____	$3 - 2 =$ ____
$5 - 2 =$ ____	$4 - 1 =$ ____	$3 - 3 =$ ____

Mixed Practice 6

1. Do these events happen in the evening/night, in the morning, or in the afternoon? Tell your teacher.

reading a bedtime story

sunrise

going swimming

2. Add. Notice the patterns.

a.	b.	c.
$2 + 0 =$ _____	$1 + 1 =$ _____	$3 + 0 =$ _____
$2 + 1 =$ _____	$1 + 2 =$ _____	$3 + 1 =$ _____
$2 + 2 =$ _____	$1 + 3 =$ _____	$3 + 2 =$ _____

3. If you put together these shapes, what new shape will you get?

a. 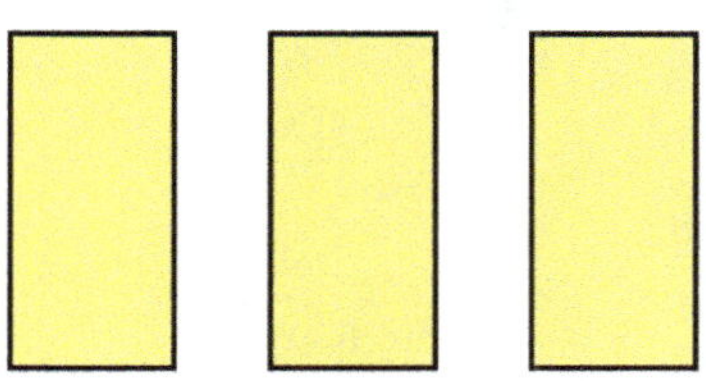

b. 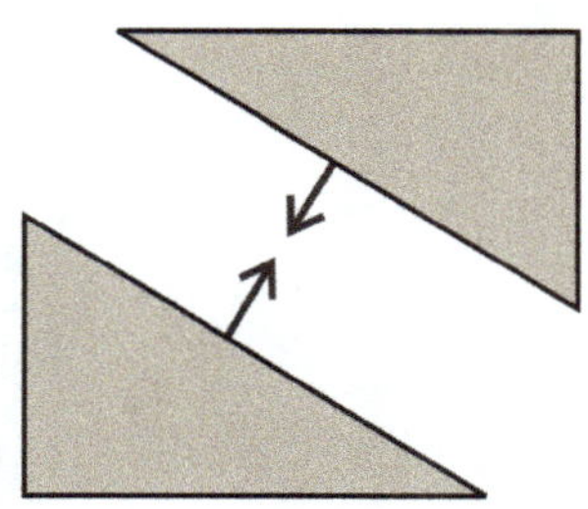

c. 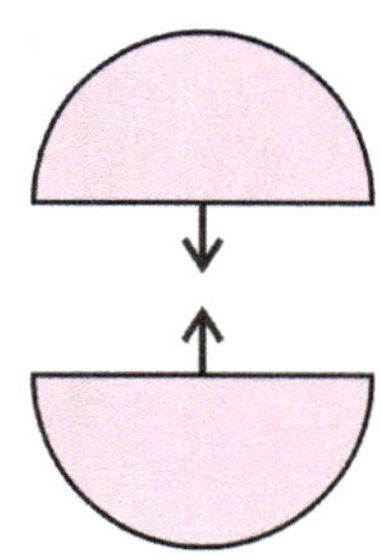

4. Circle all the words that accurately describe the animals.

small long light big

tall big long light

tall heavy small big

big long heavy small

More Practice with Numbers

1. **Activity.** You need: A 100-bead abacus.

   a. Have the child practice counting by tens from 10 to 100 using the 100-bead abacus.

   b. Ask the child to choose a number between 20 and 100 (or you give them the number), and then ask the child to count from that number to some number that you decide, so that they will count more than 10 but less than 30 steps. For example, if the child chooses 56, you could have the child count to 70.

   c. Ask the child to count from 1 to 100.

   d. Have the child use the abacus to make the following numbers:

   forty-seven   seventy-three   twenty-nine   eighty-one   thirty-five   fifty-six   ninety-four

2. Say the number word for the amount. Then draw *another* ten-bundle. Say the new amount.

a. 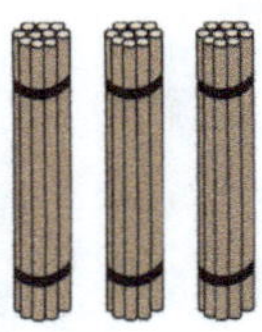	**b.** 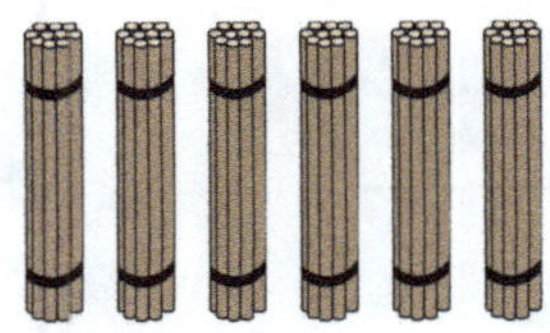	**c.** 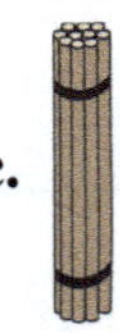

3. Match.

13	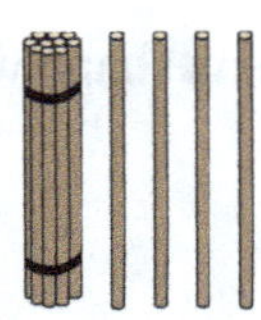
40	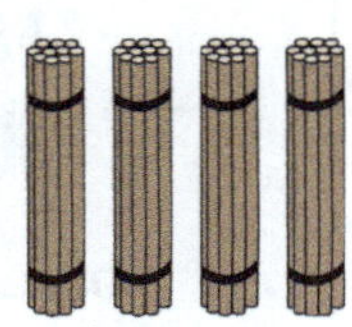
30	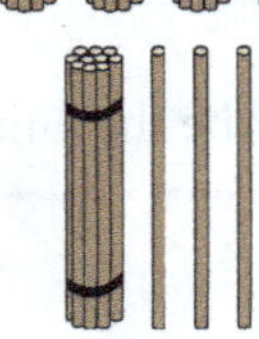
14	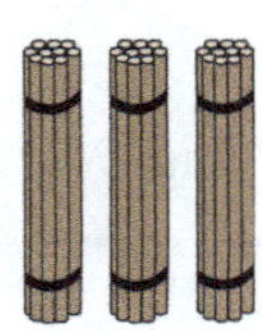

4. Draw more sticks or bundles of sticks to illustrate the given number.

a. 20 	**b.** 14 	**c.** 18 

5. Circle the specified number, in bundles of ten.

a. fifty	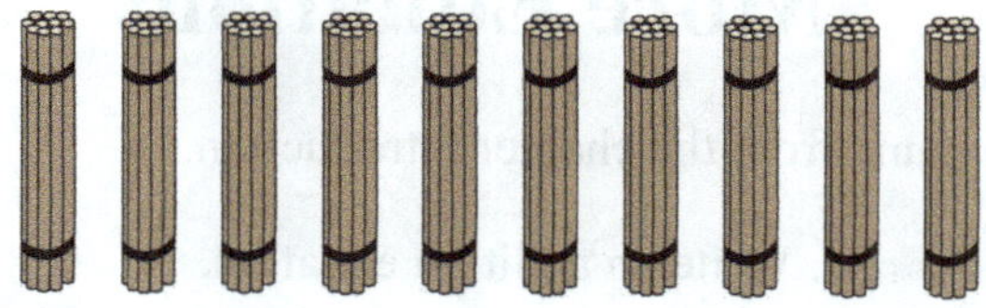
b. eighty	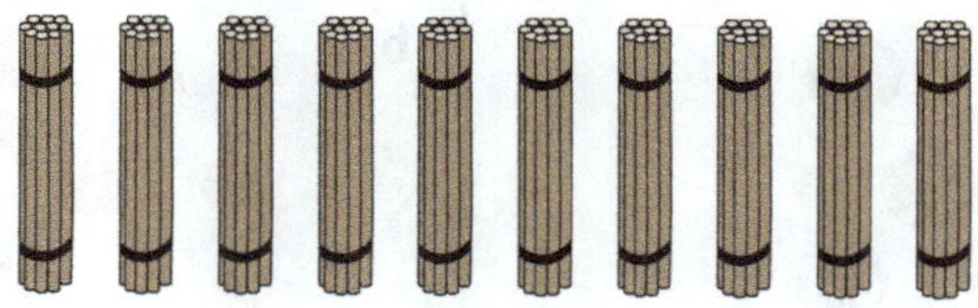
c. twenty	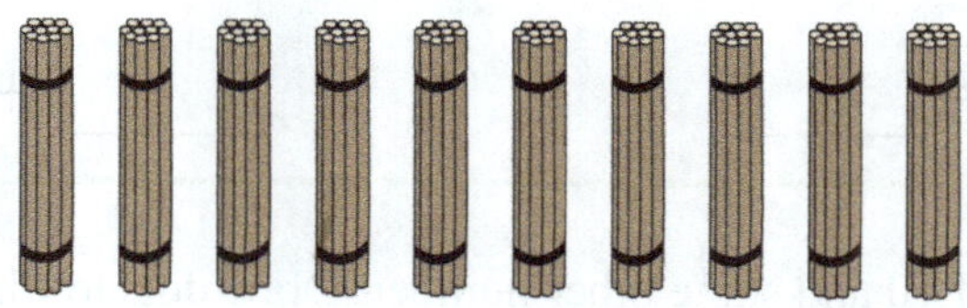

Puzzle Corner

How many hearts are there in total?

More Equations

1. Play an addition or subtraction game from the chapter introduction.

2. Circle ten objects. Then circle the rest. Write an addition equation.

a.

10 + ____ = ____

b.

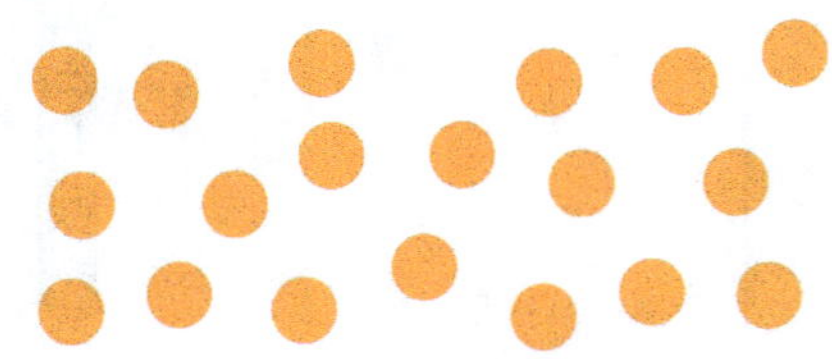

____ + ____ = ____

3. Write each number as a sum of 10 and some other number. Draw dots to illustrate the equation: draw a group of 10 dots, and some more dots.

a. 17 = ________ + ______

b. 15 = ________ + ______

4. In **a.**, draw a second group of fish so that the total is the target number. In **b.**, draw two groups of fish that make the target number. Write the addition that each picture shows. Then write a subtraction.
(Note: there are two possible additions and two possible subtractions that can be written. The child can write either one.)

a.

____ + ____ = ____

____ − ____ = ____

b.

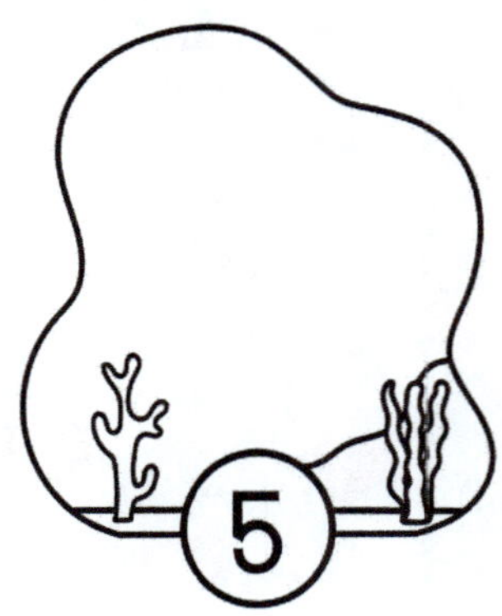

____ + ____ = ____

____ − ____ = ____

5. Add.

a. 8 + 1 = ______	**b.** 1 + 3 = ______	**c.** 2 + 3 = ______

Word Problems: More Practice

1. Solve. You can draw pictures to illustrate the problems, or act them out with objects.

a. Six birds were sitting in an oak tree.
Some were bluebirds and four were blackbirds.
How many were bluebirds?

b. Rosie baked nine pies. Two were pumpkin pies and some were apple.
How many apple pies did she bake?

2. Draw pictures to illustrate the problems. Solve.

a. Mommy wants to boil ten eggs for a picnic, but she only has six. How many more eggs does she need?

b. Jeff drew three circles, two squares, and four triangles. How many shapes did he draw in total?

3. Solve the problem.

You have this much money: 

You buy a piece of candy for 15¢.
How much money do you have left?

4. Add and subtract.

a.	**b.**	**c.**
$7 + 0 =$ ______	$4 + 2 =$ ______	$1 + 4 =$ ______
$5 - 1 =$ ______	$4 - 2 =$ ______	$2 - 2 =$ ______

Mixed Review Chapter 8

1. **Activity.** You need: An assortment of blue, red, and green objects (or some other colors). There should be a maximum of 10 objects of any one kind. Vary the number of objects in each group.

   Scatter the objects in front of the child. Ask the child to sort the objects by color into three different groups.

   After that is done, ask the child to organize the three groups by count. For example, the group with most objects could be on the left, the group with the least objects on the right, and the third group in the middle.

2. Answer your teacher.

   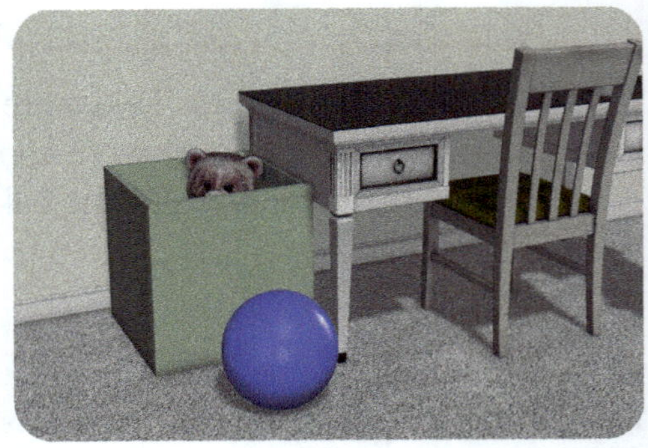

   a. In what basic shape is the green box?

   b. Where is it, in relation to the desk?

   c. In what basic shape is the blue ball?

   d. Where is it, in relation to the box?

3. Circle the group that has more things. If both have the same amount, circle both.

   a. 

   b. 

   c. 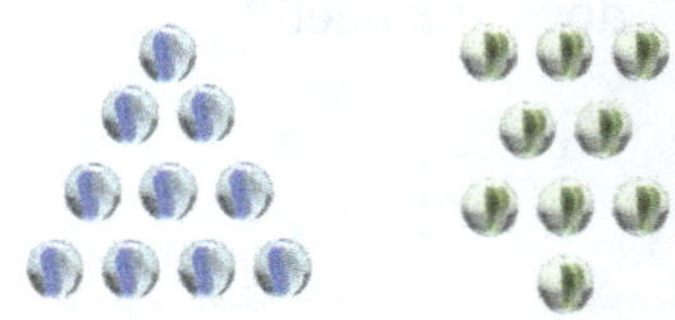

4. Write each number as a sum of 10 and some other number. Draw dots to illustrate the equation.

   a. 12 = ________ + ______

   b. 15 = ________ + ______

5. Draw pictures to illustrate the situations.

   a. A worm is longer than a pencil.

   b. A pink triangle is shorter than a red flower.

6. What is the same and what is different about these two shapes?
(Discuss with your teacher.)

7. **a.** Name each shape.

b. What is the same and what is different about these two shapes?
(Discuss with your teacher.)

8. Tell a story (or stories) about this picture that uses math and numbers.
(There is one swan, several ducks, and several fish in the picture.)

9. Color the 4th monkey **from the right**.

Circle the 5th monkey **from the left**.

Cross out the 1st monkey **from the left**.

10. Write an addition to match each illustration.

a. ________ + ________ = ________

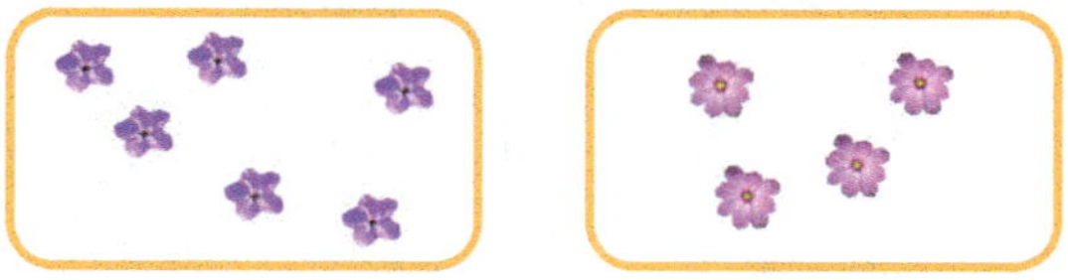

b. ________ + ________ = ________

11. You only need *one* more coin to make the target amount of money. Draw that coin (dime, nickel, or penny).

a. 15¢ 

b. 12¢ 

c. 9¢ 

d. 17¢ 

Chapter 8 Review

1. Write two subtractions to match the illustration.

_____ – _____ = _____

_____ – _____ = _____

2. Circle ten objects. Then circle the rest. Write an addition equation.

a.

_____ + _____ = _____

b.

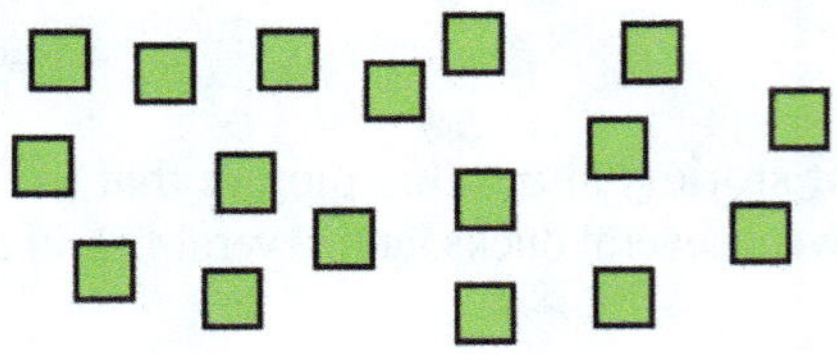

_____ + _____ = _____

3. Solve. You can draw a picture or use objects to help you.

Mrs. Kelly needs nine paintbrushes for an art class. She already has five. How many more paintbrushes does she need?

4. Write each number as a sum of 10 and some other number. Draw dots to illustrate the equation.

a. 13 = _______ + _____

b. 16 = _______ + _____

5. Add and subtract.

a.	**b.**	**c.**
5 + 0 = _____	2 + 2 = _____	9 – 1 = _____
5 – 2 = _____	4 + 2 = _____	7 – 2 = _____
3 – 2 = _____	9 + 1 = _____	5 – 2 = _____

Kindergarten End-of-Year Test

This test is fairly long, so feel free to have your child or student do it in more than one sitting. Use your judgment. The test is to be used as a diagnostic test. Thus, you may even skip those areas and concepts that you already know for sure your student has mastered.

The test does not cover every single concept that is typically included in kindergarten math, but it does cover the most important concepts and ideas. It evaluates the child's ability in the following content areas:

- counting items within 0-20, and knowing the counting sequence up to 100
- writing numbers up to 20
- comparing numbers within 0-10
- decomposing numbers within 11-19 into ten ones and some additional ones
- decomposing a number within 0-10 in several different ways
- concept of addition
- adding more to a given number in order to have 10 in total
- concept of subtraction
- basic shapes
- basic position words (under, behind, beside/next to, above, in front, etc.)

In order to start with *Math Mammoth Grade 1*, I recommend that the child gain a minimum score of 80% on this test, and that the teacher or parent review with them any content areas that are found weak. Children scoring between 70 and 80% may also go to grade 1, depending on the types of errors (careless errors or not remembering, vs. lack of understanding).

Please note that the concept of decomposing numbers within 11-19 is the least crucial aspect of this test, as it will be covered in grade 1. However, being able to *count* objects within that range is necessary before starting grade 1.

Instructions to the teacher:

Read each question to the child. The child is not required to write any words (e.g. questions 22-23, 25-28). Some questions can be done totally orally and are noted as such. Overall, the child should write the numbers in all the questions that require such unless the child has a learning disability. In such cases, use your judgment.

My suggestion for grading is below. The total is 147 points. A score of 117 points is 80%. A score of 103 points is 70%.

Question	Max. points	Student score
Counting and Writing Numbers		
1	6 points	
2	6 points	
3	4 points	
4	4 points	
	subtotal	/ 20
Comparing		
5	4 points	
6	3 points	
7	3 points	
8	6 points	
	subtotal	/ 16
Teen Numbers		
9	8 points	
10	8 points	
	subtotal	/ 16
Decomposing Numbers		
11	6 points	
12	6 points	
13	6 points	
	subtotal	/ 18

Question	Max. points	Student score
Addition		
14	6 points	
15	4 points	
16	6 points	
17	4 points	
	subtotal	/ 20
Add and Subtract		
18	6 points	
19	4 points	
20	6 points	
21	12 points	
22	6 points	
	subtotal	/ 34
Shapes and Positions		
23	8 points	
24	3 points	
25	the same: color (or being 2-D) 1 point	
	different: how many corners (or sides) 1 point	
26	3 points	
27	3 points	
28	4 points	
	subtotal	/ 23
	TOTAL	**/ 147**

Kindergarten End-of-Year Test

Counting and Writing Numbers

1. **To the teacher:** This exercise is to be done orally.

   a. Ask the student to count from 40 to 60.

   b. Ask the student to count from 87 to 100.

   c. Ask the student to count from 10 to 100 by tens.

2. Count and write the number.

a.  ____

b.  ____

c. ____

d.  ____

e.  ____

f.  ____

3. Count and write the number.

a. ____________

b. ____________

c. ____________

d. ____________

4. Group (draw a large circle around) the given number of items.

a. Make a group of eleven marbles.

b. Make a group of sixteen chicks.

c. Make a group of seventeen stars.

d. Make a group of nineteen strawberries.

Comparing

5. Circle the group that has MORE things. Circle both if they have the same number of things.

a. 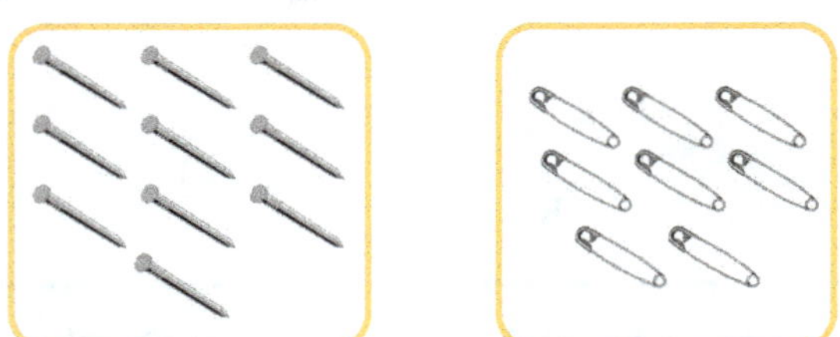

b. 

c. 

d. 

6. Circle the bigger number.

a. 9 7	**b.** 8 10	**c.** 5 4

7. Circle the smaller number.

a. 1 4	**b.** 5 6	**c.** 9 8

8. Write the number that is 1 more.

a. 8 ______	**b.** 6 ______	**c.** 3 ______
d. 12 ______	**e.** 16 ______	**f.** 19 ______

Teen Numbers

9. Each illustration has between 11 and 19 objects. Circle the ten objects. Write the total as a sum of "10 + some number".

a. 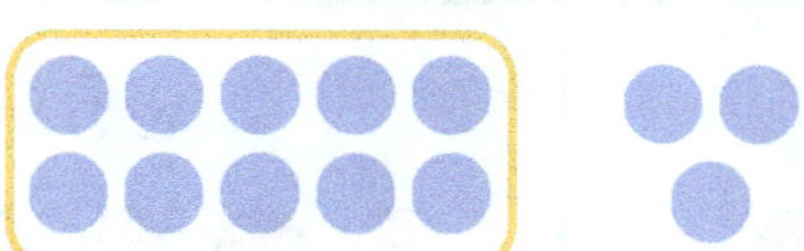

10 + __3__

__13__

b. 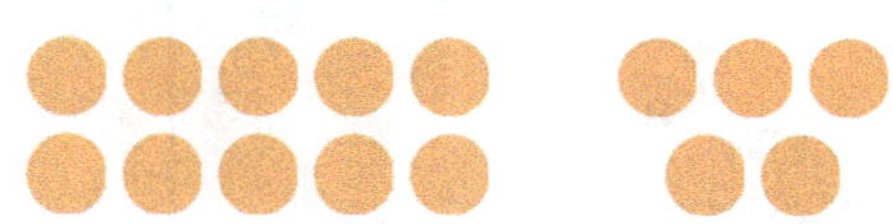

__10__ + ______

______

c. 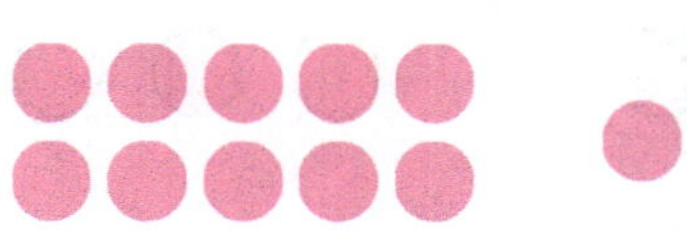

______ + ______

______

d. 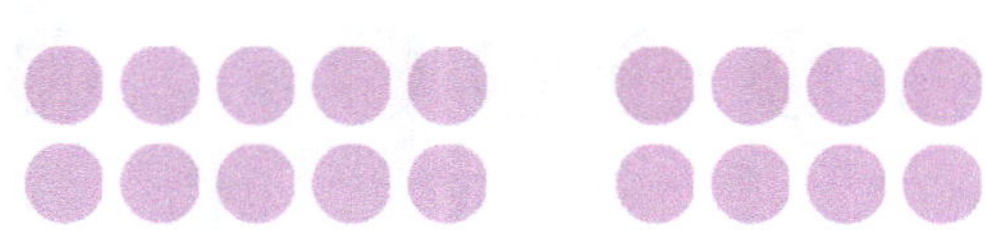

______ + ______

______

10. Write each number as a sum of 10 and some other number. Draw dots to illustrate the equation: draw a group of 10 dots, and some more dots.

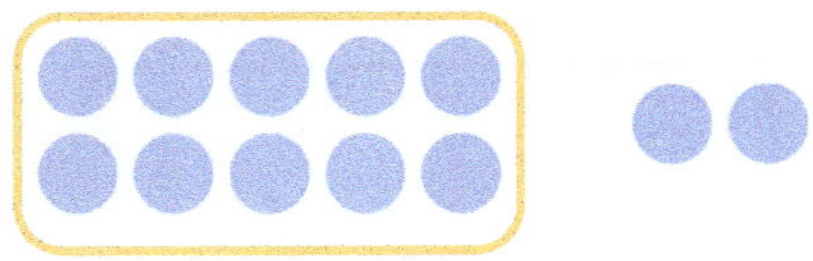 **a.** 12 = ______ + _____	**b.** 17 = ______ + _____
c. 14 = ______ + _____	**d.** 19 = ______ + _____

11. Make two groups. Write how many are in the second group.

a. 4 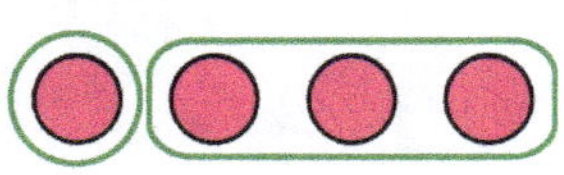 1 and __3__	**b.** 4 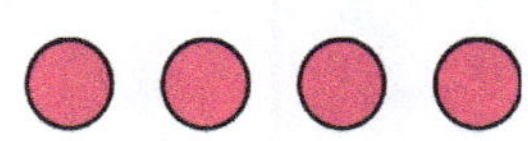 2 and _____	**c.** 5 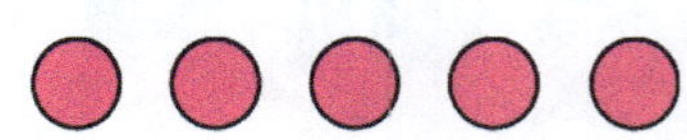 4 and _____
d. 5  1 and _____	**e.** 5  3 and _____	**f.** 5 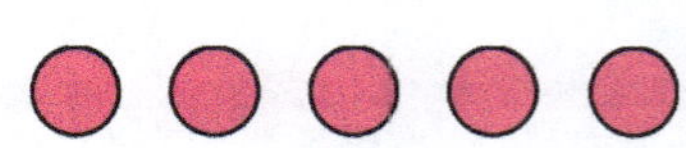 0 and _____

12. Decompose 6 and 8 into pairs in different ways. First, draw as many dots as the number shows. Then divide them into two groups, in different ways. Lastly write how many are in each group.

a. 6 _____ and _____	**b.** 6 _____ and _____	**c.** 6 _____ and _____
d. 8 _____ and _____	**e.** 8 _____ and _____	**f.** 8 _____ and _____

13. The number at the top is the total. Draw the missing dots on the face of the blank dice. Write on the lines how many dots are on the face of each dice.

a. 3

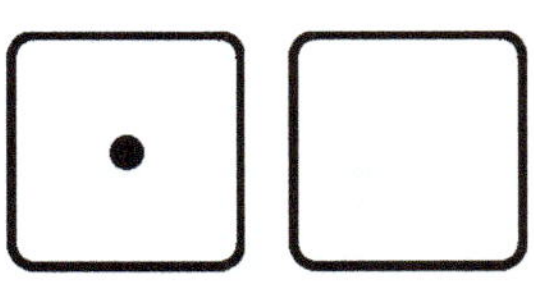

_____ and _____

b. 6

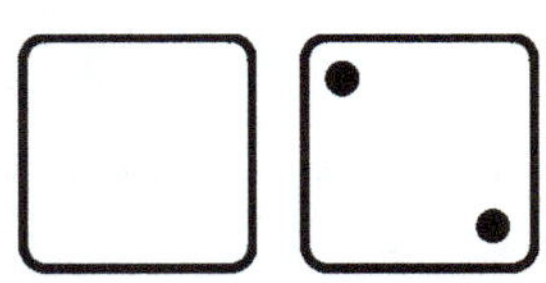

_____ and _____

c. 5

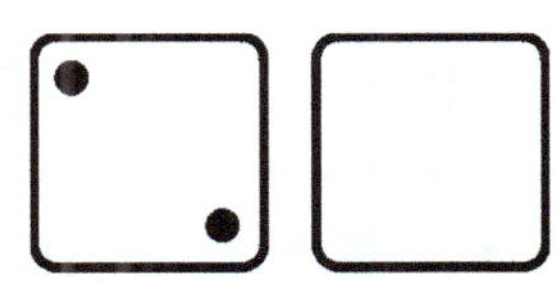

_____ and _____

d. 4

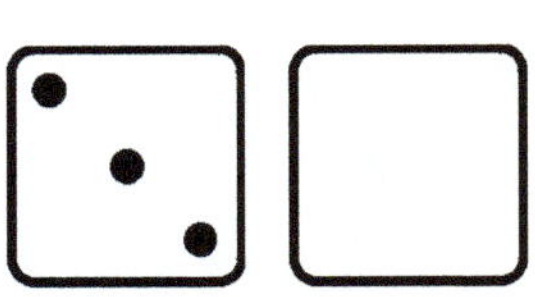

_____ and _____

e. 6

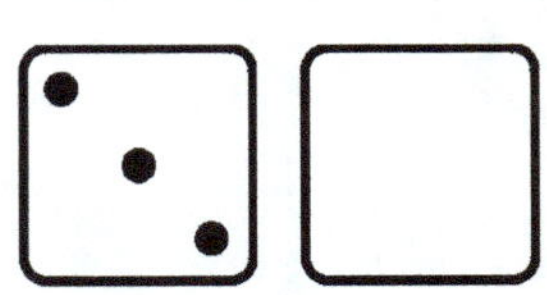

_____ and _____

f. 5

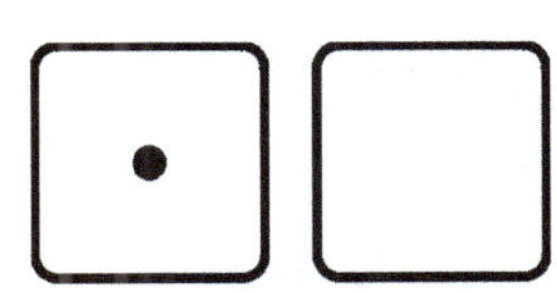

_____ and _____

Addition

14. Write the number of things in each group. Add. Read the addition sentences aloud using "plus" and "equals".

a. 

1 + 3 = ☐

b. 

_____ + _____ = ☐

c. 

_____ + _____ = ☐

d. 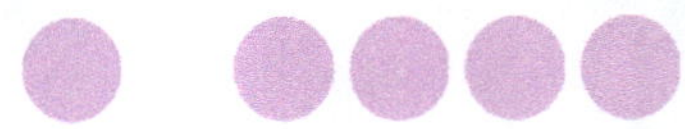

_____ + _____ = ☐

e. 

_____ + _____ = ☐

f. 

_____ + _____ = ☐

15. Add with zero.

a. 

_____ + _____ = _____

b. 

_____ + _____ = _____

c. 

_____ + _____ = _____

d. ☐ ☐

_____ + _____ = _____

16. Add. You can draw sticks, circles, or other pictures to help you.

a. 4 + 3 = _____	**b.** 5 + 5 = _____	**c.** 7 + 1 = _____
d. 9 + 0 = _____	**e.** 3 + 6 = _____	**f.** 3 + 4 = _____

17. Draw another group of sticks so that you will get 10 in total. Write the numbers in each group.

a. 

_____ and _____

b. 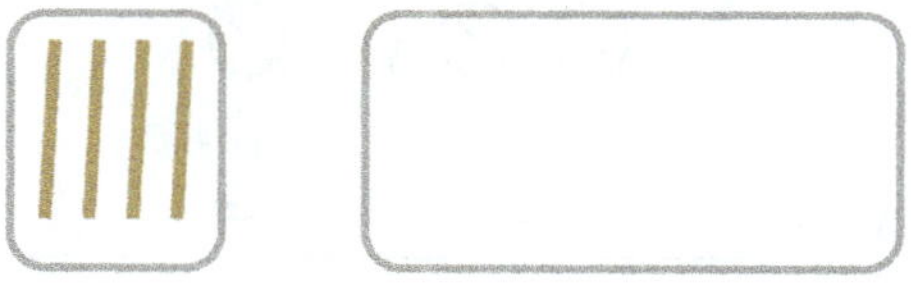

_____ and _____

c. 

_____ and _____

d. 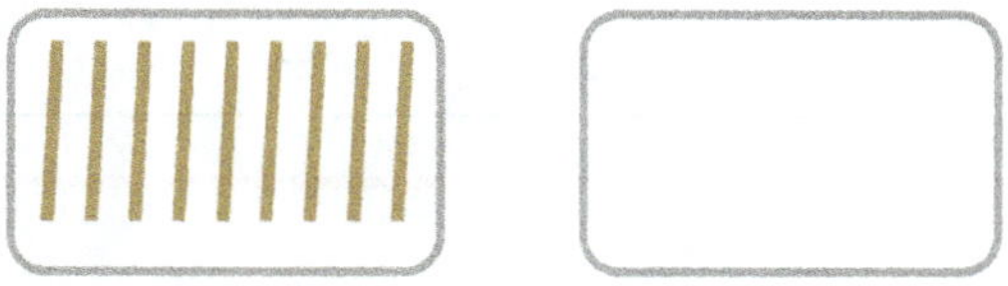

_____ and _____

Add and Subtract

18. Cross out objects. How many are left?

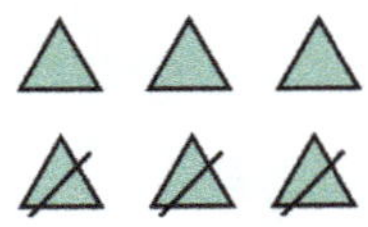

a. 6 – 3 = __3__

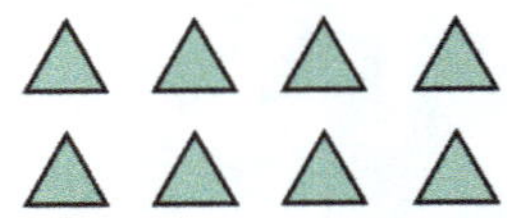

b. 8 – 6 = ______

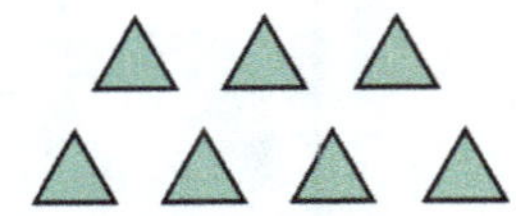

c. 7 – 3 = ______

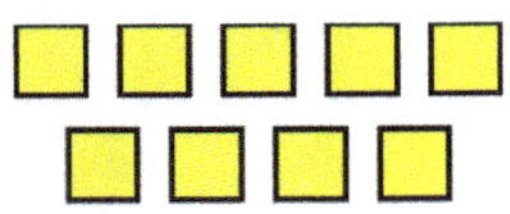

d. 9 – 3 = ______

e. 8 – 4 = ______

f. 5 – 5 = ______

19. Write a subtraction sentence to match each picture.

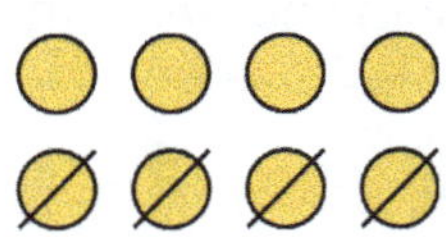

a. ______ – ______ = ______

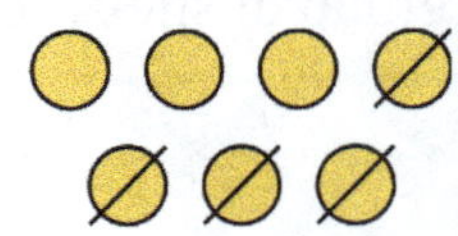

b. ______ – ______ = ______

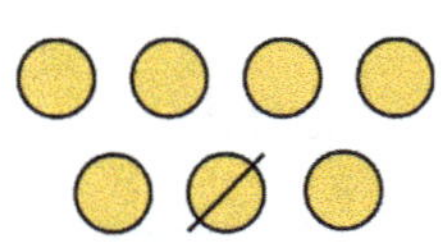

c. ______ – ______ = ______

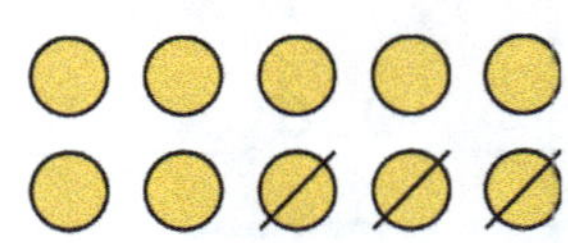

d. ______ – ______ = ______

20. Subtract. You can draw sticks or circles to to help you.

a. 8 – 3 = ______

b. 5 – 1 = ______

c. 10 – 7 = ______

d. 7 – 2 = ______

e. 6 – 3 = ______

f. 9 – 7 = ______

21. Add and subtract.

a. $4 + 1 =$ ____ $2 + 2 =$ ____	**b.** $2 + 3 =$ ____ $1 + 1 =$ ____	**c** $3 + 0 =$ ____ $0 + 5 =$ ____
d. $2 - 2 =$ ____ $5 - 3 =$ ____	**e.** $3 - 2 =$ ____ $4 - 3 =$ ____	**f.** $4 - 0 =$ ____ $3 - 1 =$ ____

22. Solve. You can draw something to help you, or act it out. (Can be done orally.)

a. Henry saw three cookies on one plate and five cookies on another. How many cookies in total did Henry see?

b. There were nine birds in a tree. Then two of them flew away. How many birds are still in the tree?

c. Lily drew five bunnies sleeping. Then she drew two more. How many bunnies did Lily draw in all?

Shapes and Positions

23. Name each shape. (To be done orally.)

a. 

b. 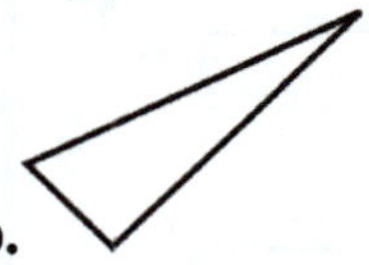

c. 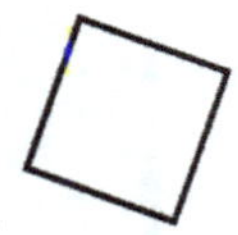

d. 

e. 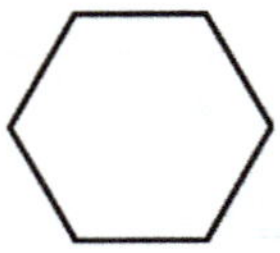

f. 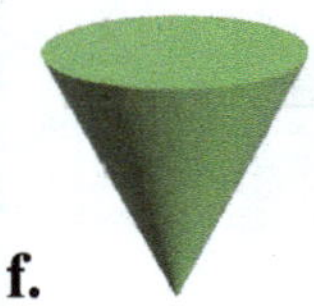

g. 

h. 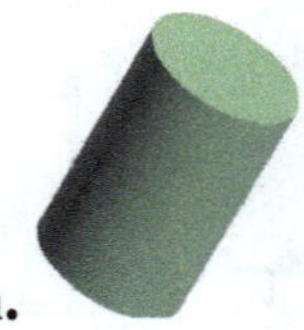

24. Draw here a triangle, circle, and rectangle. (An approximate sketch is sufficient.)

a. Draw a triangle.	**b.** Draw a circle.	**c.** Draw a rectangle.

25. What is the same and what is different about these two shapes?
(To be done orally.)

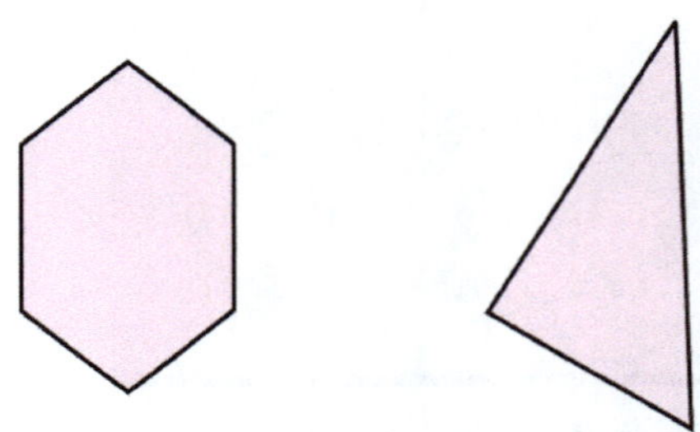

26. If you put together these shapes, what new shape will you get?
(To be done orally, you may cut out the shapes.)

a.

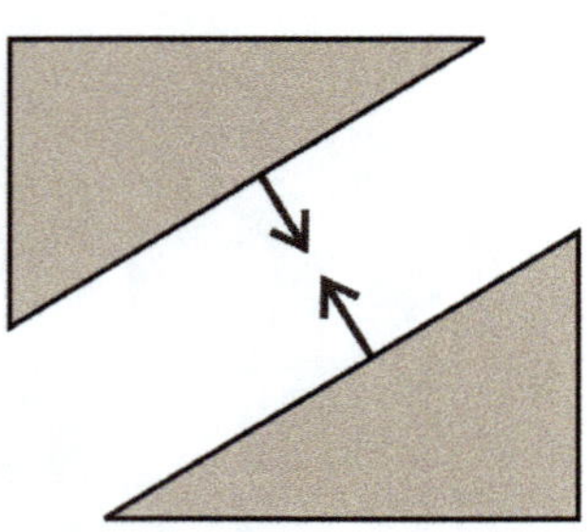

b.

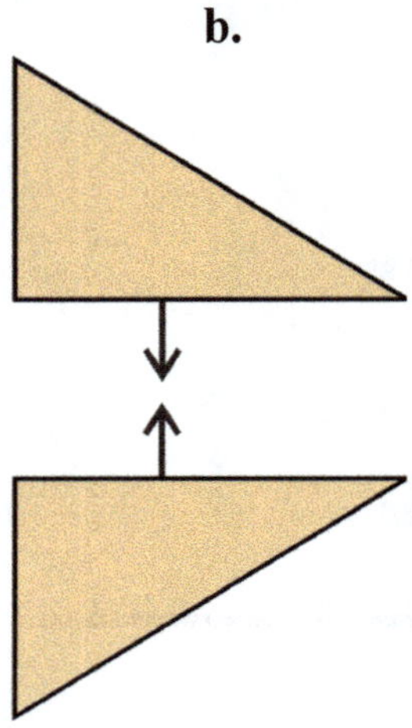

c.

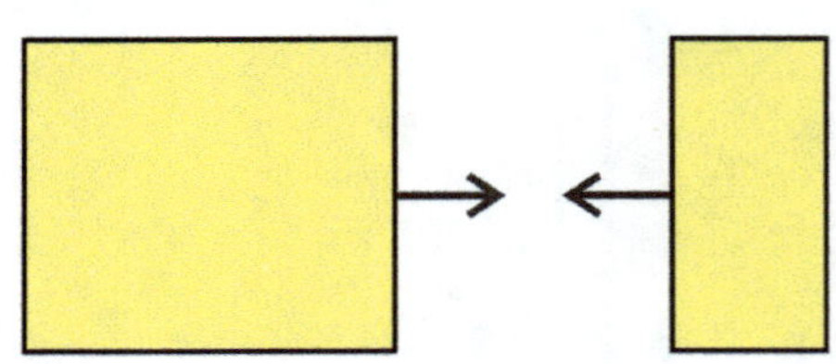

27. Are the shapes flat/two-dimensional (2-D) or solids/three-dimensional (3-D)?
(Can be done orally, or the child can draw a line from each shape to the correct answer.)

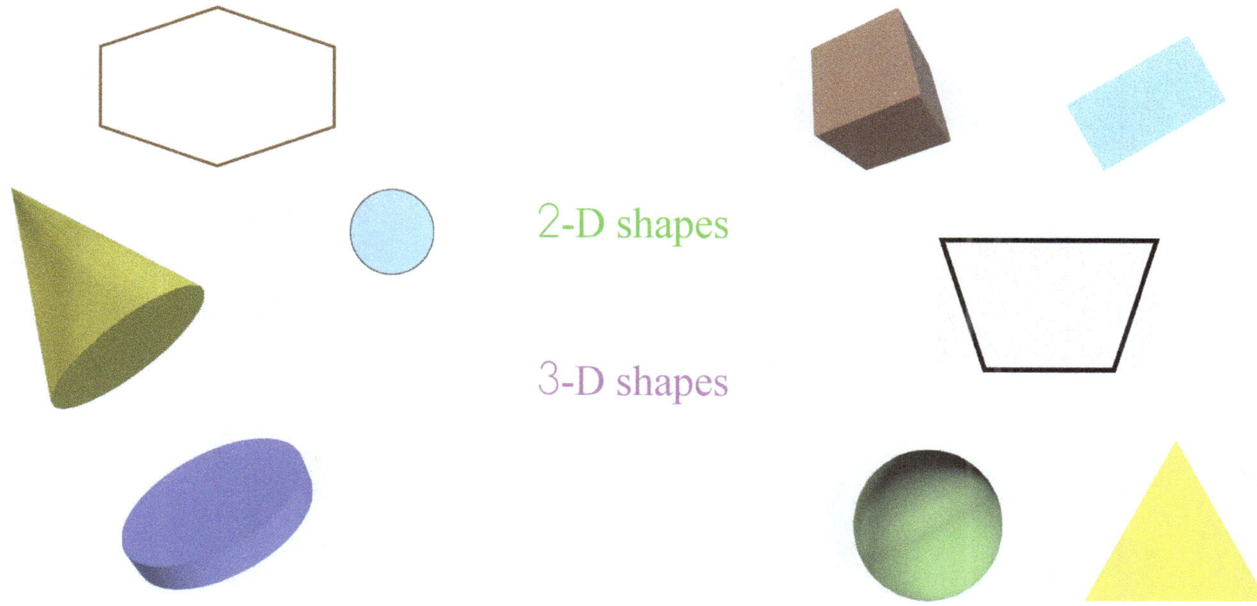

28. Where is the cat? Describe the relative positions of objects. (To be done orally.)

 **a.** The cat is ____________________ the box.	 **b.** The table is ____________________ the sofa.

29. Answer (orally).

a. Where is the cat (in relation to the couch)?

b. In what basic shape is the party hat?

Where is it (in relation to the couch)?

c. In what basic shape is the die (or dice)?

Where is it (in relation to the couch)?

www.ingramcontent.com/pod-product-compliance
Lightning Source LLC
Chambersburg PA
CBHW061221240926
51846CB00006B/1301
9781954379503